MW00709538

Successful
Industrial
Experimentation

Successful Industrial Experimentation

Brett Kyle

VCH

Brett Kyle
581 Barons Court
Burlington, Ontario
Canada, L7R 4E4

This book is printed on acid-free paper. ♾

Library of Congress Cataloging-in-Publication Data

Kyle, Brett.   Successful industrial experimentation / by Brett Kyle.
　　p.　　cm.　　Includes bibliographical references and index.
　ISBN 1-56081-050-5 (alk. paper)
　　1. Experimental design.　　2. Research, Industrial—Statistical methods.　 I. Title.
T57.37.K95　1995
607'.2—dc20   95-6060
                                                                                                         CIP

© 1995 VCH Publishers, Inc.

Printed in the United States of America

ISBN 1-56081-050-5 VCH Publishers, Inc.

Printing History:
10　9　8　7　6　5　4　3　2　1

Published jointly by

VCH Publishers, Inc.　　　　VCH Verlagsgesellschaft mbH　　VCH Publishers (UK) Ltd.
220 East 23rd Street　　　　　P.O. Box 10 11 61　　　　　　　8 Wellington Court
New York, New York 10010　　69451 Weinheim, Germany　　　Cambridge CB1 1HZ
　　　　　　　　　　　　　　　　　　　　　　　　　　　　　United Kingdom

To my wife, Teresa, and children, Sean and Rachel

Preface

This text has been developed as an introductory guide to process and product improvement through the use of simple quality improvement techniques and experimental design methodology. The fundamentals of a sound experimental approach to problem solving that incorporates valid statistical analysis is stressed. The success of an experiment depends as much on the amount of planning and preparation that is done as it does on the theoretical background behind the statistics and designed experiments. Many experiments can be run and analyzed with a basic understanding of descriptive and inferential statistics, with the most successful being the ones that have had sufficient time devoted to up-front planning.

All techniques presented in this text have been applied in a variety of industrial situations by individuals with varying academic backgrounds. Key instruction sets have been clearly outlined to assist in the problem-solving process and a cookbook approach has been adopted for illustrating the various techniques available. This text is particularly beneficial to those individuals who are interested in performing experiments but have no inclination to become a statistician.

Users of this text are cautioned about the possibility of misapplication of these techniques. If there is any doubt about the design being contemplated for a given experimental setup, consultation with a qualified statistician is recommended. In fact, the text will be an ideal medium for introducing some of the basic terminology required for conversing with a statistician.

This text is divided into four main sections. The first section introduces some philosophy and then focuses the reader's attention on some of the simple quality tools recommended by quality experts to help an experimenter prepare for an experiment. Specific topics covered include (1) data collection, (2) flow diagrams, (3) Pareto analysis, and (4) cause and effect diagrams.

Following the introduction of the simple tools, a description of some of the fundamental statistical concepts such as central measures, variability, probability distributions, and analysis of variance (ANOVA) is provided. The calculation and interpretation of the mean, variance, and standard deviation are presented as part of this second section.

The concept of screening designs is discussed next. Orthogonal arrays will be used almost exclusively in this text for screening designs. Although these designs are being used, many other types of screening designs (e.g., Plackett–Burman, Greco-Latin Squares, and Hadamard) are available to the experimenter. However, to minimize the complexity of the text only orthogonal arrays will be discussed in any detail. Only one data analysis technique, the analysis of variance or ANOVA, will be discussed. Information on other designs and data analysis techniques can be obtained by consulting the reference texts mentioned in Appendix E.

Finally, the reader will be provided with tips on which arrays to use, how to deal with interactions, variable randomization, factor selection, use of noise factors, and how to choose good response variables.

It is expected that readers will obtain sufficient knowledge from this text to incorporate these techniques into their own job function and to be able to conduct and interpret a statistically designed experiment.

Brett Kyle
Burlington, Ontario
Canada
August, 1995

Acknowledgments

The author is grateful to the companies that allowed the experiments, conducted on their behalf, to be used as examples for this text. Appreciation is also extended to those participants in the various workshops presented by the author. Many of their suggestions for improving the presentation of experimental design techniques provided the basis for approaches used in this text.

In addition, this text could not have been possible without the understanding of the author's wife and children during the preparation of this manuscript. Their patience during the many days (and nights) of writing is greatly appreciated. Thanks are also due to Ms. P. Kyle, Ms. L. Kane, Mr. S. Haynes, and Mr. A. Zahavich who unselfishly donated their time to proofread the manuscript and to evaluate its readability, grammatical accuracy, and statistical correctness.

Contents

1. Introduction   1
1.1   Why Conduct Experiments?   1
1.2   Stages of Experimentation   4
      Stage 1: Selection of a Design Team   4
      Stage 2: Preparation for Experiment   6
      Stage 3: Conduct Trial   7
      Stage 4: Data Analysis   8
      Stage 5: Publicize Results   8

2. Getting Started   9
2.1   Flow Diagrams   9
2.2   Construction of a Flow Diagram   10
      Step 1: Assembling a Team   12
      Step 2: Identification of the Main Flow Path   12
      Step 3: Addition of Secondary Loops   14
      Step 4: Check Consistency   14

3. Data Collection   15
3.1   Check Sheets   15
3.2   Types of Check Sheets   15
3.3   Construction of a Check Sheet   16

Step 1:   Selection of the Design Team   16
Step 2:   Identifying Collection Points   18
Step 3:   Duration of Data Collection   18
Step 4:   Collecting Data   18
Step 5:   Data Review   19
3.4   Disappearing Income   20

4. Pareto Charts   25
4.1   Pareto Principle   25
4.2   Construction of a Pareto Chart   27
Step 1:   Categorize Data   28
Step 2:   Formatting   28
Step 3:   Calculation of Relative and Cumulative
          Frequency   29
Step 4:   Plot Data   30

5. Cause and Effect Diagrams   33
5.1   Introduction   33
5.2   Types of Diagrams   35
5.3   Construction of Cause and Effect Diagrams   37

6. Basic Statistics   43
6.1   Introduction   43
6.2   Descriptive Statistics   44
6.3   Samples and Populations   44
6.4   Average   45
6.5   Mean   45
6.6   Median   46
6.7   Variation   48
6.8   Standard Deviation and Variances   49
6.9   Normal Distribution   53
6.10 Hypothesis Formulation   54
6.11 Error and Risk   56
6.12 F Statistic   57

7. Analysis of Variance   61
7.1   One-Way Analysis of Variance   61
7.2   Changing Fuel Efficiency   61
7.3   Types of Variation   63
7.4   Variance   64
7.5   Sum of Squares   64
7.6   Degrees of Freedom   64
7.7   Generation of a One-Way Analysis of Variance Table   65

Step 1:   Determination of Squares   66
Step 2:   Calculation of the Correction Factor   67
Step 3:   Calculation of the Sum of Squares   67
Step 4:   Calculation of the Degrees of Freedom   68
Step 5:   Generation of ANOVA Table   68
7.8   Two-Way Analysis of Variance   70
Step 1:   Determination of Squares   71
Step 2:   Calculation of the Correction Factor   73
Step 3:   Calculation of the Sum of Squares   73
Step 4:   Calculation of the Degrees of Freedom   74
Step 5:   Generation of ANOVA Table   74
7.9   Determination of F Statistic   75

8. Screening Experiments   77

8.1   What are Screening Experiments?   77
8.2   Orthogonal Arrays   77
8.3   Confounding   78
8.4   Resolution   82
8.5   Selecting a Design   82
8.6   Selecting Response Variables   83
8.7   Assigning Factors to a Design   84
8.8   Choosing Factor Ranges   85
8.9   Determining the Reasonableness of the Ranges
        Selected   85
8.10   Sample Size   86
8.11   Determination of Sample Sizes   86

9. ANOVA for Screening Experiments   89

9.1   Case Study: Batting Average   89
9.2   Calculation of the Level Sums   91
9.3   Level Sums Table   93
9.4   Sums of Squares   94
9.5   Significance Plot   95
9.6   Pooled Error ANOVA Table   96
9.7   Variance   97
9.8   Calculation of the F Statistic   97
9.9   Expected Sums of Squares and Percent Contribution   98
9.10   Interpretation of ANOVA Table   99
9.11   Estimation of Best Run Conditions   100
9.12   Confidence Intervals   101
9.13   Confirmation Experiment   102

Appendix A. Glossary   103

**Appendix B. Common Two- and Three-Level Orthogonal
                         Arrays   107**

    L_4 107
    L_8 108
    L_9 108
    L_{12} 109
    L_{16} 109
    L_{18} 110
    L_{27} 110
    L_{32} 112

**Appendix C. Confounding Patterns for Common
                         Orthogonal Arrays   115**

    L_8 115
    L_{16} 117
    L_{32} 122

Appendix D. Tables   127

    F Distribution   127
    t Distribution   130
    Random Numbers   131
    Values for t_α and t_β at Selected Degrees of Freedom   131

Appendix E. References   133

Index   137

1

Introduction

"Here is Edward bear coming downstairs now, bump, bump, bump, on the back of his head, behind Christopher Robin. It is as far as he knows the only way of coming downstairs, but sometimes he feels that there is another way, if only he could stop bumping for a moment and think of it. And then he feels that perhaps there isn't. Anyhow here he is at the bottom."

A. A. Milne's, "Winnie the Pooh." Reprinted with permission.

"He uses statistics as a drunken man uses lamp posts—for support rather than illumination."

Andrew Lang, 1884–1912, Scottish author

1.1 Why Conduct Experiments?

Conducting experiments is an everyday occurrence in modern life. An experiment might be conducted to

- Determine the feasibility of a new idea or approach;
- Screen potential candidates for substitution into an existing product;
- Develop an entirely new product;
- Determine how production variables affect product and process quality;
- Optimize a product or process;
- Solve a problem;
- Reduce costs.

Where does one start to tackle some of these tasks? the dynamic and highly

1

competitive nature of today's market demand that problems be solved effectively and efficiently. Generally speaking, financial and human resources are limited within an organization. Therefore, the strategy used to investigate an opportunity must be simple, easy to implement, understandable by all involved, and, of course, cost effective.

The foundations of an experimental program must be well understood and have the full commitment of almost everyone in the organization. This includes shop floor personnel as well as top level managers. If the top level managers are not committed to the program, the support needed by the operators and production engineers to successfully carry out the necessary experimentation will not be there.

Once employees are involved in the developmental process, knowledge of how the process operates and what it takes to get things done will increase. Key elements to this program would be trust and technical know-how. Attention will be focused on a particular opportunity rather than on individual bias. Decisions will begin to be made using data. Using data to gain commitment from co-workers for a particular point of view is always more productive. Being meticulous in the planning of each experiment will increase the probability of successfully completing a project. It is important to become familiar with all the techniques described in this text.

Consider the process of cooking a meal and the many alternative ways to do so. Familiarity with the recipe, availability of particular ingredients, cooking method, and even the type of pot used will all influence the final outcome. What if there was a taste difference each time the meal was prepared? What contributed to the inconsistency of the taste? Could it be the addition or substitution of an ingredient, too much or too little of one ingredient, or was the food overcooked. Could it possibly be something else that is not under the cook's control.

To determine what caused the change in taste, a series of trials, either statistically or randomly organized, could be conducted. Most commonly a random approach would be taken to this problem by initially changing one process factor at a time while all others factors are kept constant. This process would be repeated until the inconsistency in the taste was overcome to the satisfaction of the cook. The *change-one-factor-at-a-time* approach is shown graphically in Figure 1.1.

As an alternative, a statistically oriented technique could be used to find the cause of taste differences. This statistical approach is commonly referred to as designed experimentation, statistical experimental design, design of experiments, DoX, or DoE. This technique was developed in the 1920s by the renowned British statistician Sir R. A. Fisher. His motive was to improve the efficiency of the experiments he was conducting and to have a more precise method for interpreting his experimental data. A good guesser may beat a statistically designed approach occasionally, but in the long run the statistical approach will be better.

A classic example[1] of what designed experimentation can accomplish is illustrated in a 1953 experiment conducted by Ina Seito (now called INAX Corp.), a large

---

[1] Taguchi, G., Wu, Y. "Introduction to Off-line Quality Control." Central Japan Quality Control Association, Meieki Nakamura-Ku Magaya, Japan, 1979.

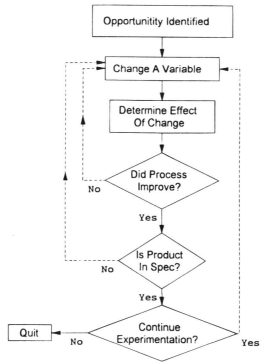

Figure 1.1. "Change-one-factor-at-a-time" problem-solving approach.

Japanese producer of ceramic tiles. The Ina Seito company had just purchased a multimillion dollar ceramic kiln. Although the kiln was new, production personnel found severe dimensional variation throughout each load of tiles coming out of the kiln. Dimensions of the tiles on the outside were found to have greater tile to tile size variation than inner dimensions. Production staff felt this variation in tile dimensions was primarily caused by an uneven temperature distribution in the kiln.

The obvious and traditional approach would have been to modify the kiln design to improve the temperature distribution. However, modification costs were estimated to be in the neighborhood of half a million dollars and an additional expenditure of this magnitude was unthinkable at the time. Since tremendous losses were being borne by the company the problem had to be solved quickly. Instead of succumbing to stereotyped problem-solving practices and modifying the design of the kiln, a nonconventional approach (at that time) was taken by the production staff and a designed experiment was conducted.

Representatives from throughout the company considered to be knowledgeable in the tile process were assembled into a team. They were allowed to brainstorm a list of variables that could potentially be contributing to the tile dimension problem. After compiling this list of variables, the ones considered the most likely to contribute to the variation in tile dimensions were then selected and finally incorporated into a designed experiment.

Appropriate statistical analysis of the experimental data identified both the significant contributors to tile variation and suggested appropriate settings for these variables. After setting up the process according to the results of this one experiment, the number of defective tiles decreased immediately to 1% from the initial 30%. In addition, an increase in the percentage of the cheapest raw material, from 1 to 5%, was found to improve product quality and decrease part-to-part variation.

In the end, superior quality tiles were produced by Ina Seito at a much cheaper price and with decreased tile-to-tile variation, all without having to modify the kiln design. This example illustrates how the application of statistical methods and simple quality tools results in a potent experimental strategy.

1.2 Stages of Experimentation

To achieve similar results to those of Ina Sieto an organization must develop a quality-oriented experimental strategy. The *stages of experimentation,* as discussed below, are one such strategy that can be used as a guideline to developing a system that works within the organizational structure of a company. An introduction to the stages of experimentation follows and they are summarized in the flow diagram illustrated in Figure 1.2.

Stage 1: Selection of a Design Team

Assuming a broad objective or a problem has been defined, the experimental process begins by selection of a design team. This team typically consists of a broad base of people with a variety of backgrounds and experiences from both inside and, if practical, outside the organization. Each organization is set up differently. Therefore, what may be acceptable to one company may not work well in another. The list provided below can be used as a guideline for selecting members for a design team.

Suggested Design Team Members

Production engineering
Management
Production foremen
Customers (where appropriate)
Research and development
Sales and marketing
Process operators (always!)
Industry consultants

A diverse design team should increase the probability of success. Each person affected by a problem or, more appropriately, an opportunity tends to have a different viewpoint of its exact nature and how it should be approached. Their ideas will be based on individual experience and training. Each one will bring one of the pieces

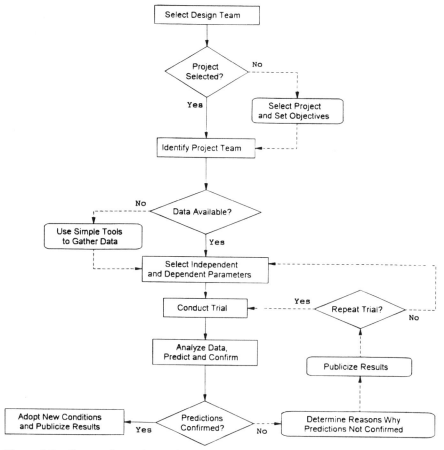

Figure 1.2. Stages of experimentation.

of the puzzle to the table. The synergy that results from getting a diverse group of people involved in a team will usually advance the problem-solving process.

Input should be encouraged from all the various line functions within an organization for another very important reason. Often it is necessary to have the support and the understanding of more than one function to be able to complete a project, in other words, each function affected by the project should buy into the program to maximize the likelihood of success. When one of the functions within an organization does not participate fully it could result in a misunderstanding of the program objectives or important factors being overlooked. This could result in an overall reduction of the effectiveness of the proposed solution.

Opportunity Identification

After deciding who will participate as members, the core design team is assembled and the various objectives of the program are defined. The key to having a successful outcome depends on determining precisely the internal and/or external needs of the customer. If there are insufficient data to determine this information or all members are not equally knowledgeable, background data must be collected and presented to the group. A number of simple tools can be employed to improve the knowledge base of the design team. The most commonly used tools are

Teams (brainstorming)
Control charts
Pareto charts
Experimental design
Data collection
Flow diagrams
Cause and effect diagrams

Each tool helps improve the understanding of how a process or product behaves. Variables important to product quality are effectively identified. For these tools to be effective, input is required from all functions, not just from one or two people. Therefore, a healthy working relationship begins to develop between the various functions in the organization. As each tool is mastered an increased sensitivity to quality will be formed and gains in productivity and quality will likely follow.

Stage 2: Preparation for Experiment

Making initial preparations for the experiment is the next responsibility of the design team. A decision on how to judge the success of the project is also required as well as choosing quality characteristics (i.e., dependent responses) and ways of measuring those parameters selected. The most probable variables that influence the project objectives or quality characteristic(s) are typically chosen from a cause and effect diagram. The team must then select an experimental design (e.g., screening design or 2 level factorial) suitable for the number of variables chosen and assign the variables to the design. Appropriate ranges for each variable must also be chosen and added to the design.

Table of Reasonableness

Next in this process is the construction of a *table of reasonableness.* This table establishes the practicality of the variables and ranges chosen for study and highlights unreasonable combinations of conditions before being carried out in the plant or laboratory. When unreasonable conditions (e.g., safety, cost, poor quality product, and time) are found, the design team must rethink its choices for variable and/or ranges for each variable. Settings are then adjusted accordingly; another table of reasonableness is constructed and the process is repeated. This is a key step in the design strategy and should not be avoided. Before proceeding with the trial the experimental plans should be published internally for review by other employees.

Pretrial Preparation

A pretrial preparation flow diagram, such as illustrated in Figure 1.3, can be used to ensure that all resources required are procured and commitment from all participants, such as material suppliers, analytical services, quality control, and production, are received so the trial will run smoothly.

Stage 3: Conduct Trial

Once the table of reasonableness has been approved and all necessary preparations have been made the experimental trials are conducted. The run order of the experiments is normally randomized to minimize the effect of experimental noise. Those experimental combinations considered to be the most difficult or the most likely to cause problems should be run first. This will verify the table of reasonableness and ensure that it is possible to run all other conditions. If these conditions can be run without great difficulty, the remaining trials in the experiment would be completed. Quality characteristics are then measured using a suitable measurement system.

Stage 4: Data Analysis

Test results are tabulated and analyzed using an appropriate technique and the results are made available to the design team. Using the results from the experiment, an

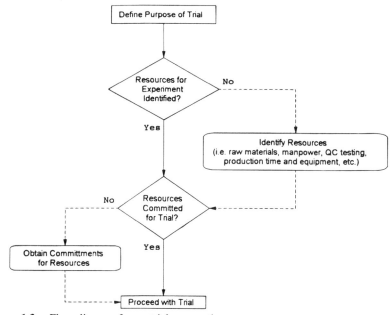

Figure 1.3. Flow diagram for pretrial preparation.

expected optimum value for the quality characteristic is calculated mathematically and another trial is run using these conditions. This is called a confirmation trial. If predictions are confirmed, the new settings can be adopted and the next set of trials organized. If predictions are not confirmed the cause of the nonconformance should be identified and a suitable solution should be implemented.

Stage 5: Publicize Results

When the experiment has been completed the results should be presented to participants in the experiment and a copy of the report should be made available to anyone who may be interested in the results.

CHAPTER

2

Getting Started

"Tell me and I will listen. Show me and I will learn. Involve me and I will understand."

A Chinese Proverb

2.1 Flow Diagrams

A flow diagram is a graphic representation of the natural progression through a *process,* where a process is defined as a system that requires an input to get started, a mechanism to manipulate the input to keep things running, and an output of some kind.

Flow diagrams can be an extremely useful diagnostic tool. For example, they can

- Be a schematic of how the "process" is perceived to be operating;
- Identify redundancies and inefficiencies in a process;
- Show employees how their function fits into the overall process;
- Focus attention on the relationship of the various functions and how quality problems in one area can affect another person's job function;
- Identify the internal customers[1] within a given process.

---

[1] An internal customer is an individual or group within a person's organization that requires a product or other types of work from another group or individual within the organization before they can complete their own job.

9

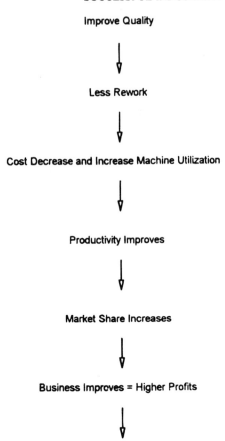

Figure 2.1.   Flow diagram using only words.

For example, a flow diagram can be used with equal effectiveness to describe how to solve problems, how to adjust the timing in a car, how to get up in the morning, how to invoice customers, or how to bake a cake. Flow diagrams can be illustrated in several different ways. They can be made up of words as shown in Figure 2.1, of engineering or programming symbols as shown in Figure 2.2, or even some combination of both of these methods.

The flow diagram constructed must be understandable. If those who use the diagram can follow the steps outlined and it is representative of the process, it has accomplished its function.

2.2 Construction of a Flow Diagram

There are four basic steps in the construction of a flow diagram:

- Assembling of a team;
- Identification of the main flow path;

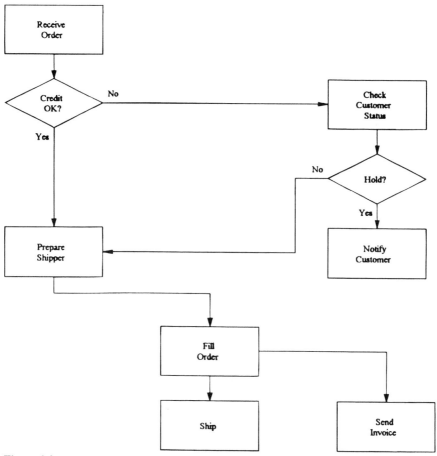

Figure 2.2. Flow diagram using symbols.

- Addition of secondary loops;
- Check consistency.

When constructing a flow diagram the following guidelines can be used:

- The flow diagram should be constructed on a large surface where it is clearly visible to everyone. This usually means taping several smaller pieces of paper together as they become filled with information;
- Make sure sufficient time is available for completing a flow diagram. Schedule extra time as it can take longer than originally expected;
- Make certain that questions are asked while constructing the flow diagram. For example, What happens first? What happens next? Does this step have to be done? What happens if the part is unacceptable at this stage? How do I know whether this part is acceptable? Where does the raw material come from for this operation? How are changes made in the process?

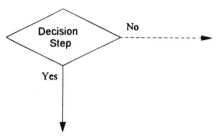

Figure 2.3.   Illustration of a decision step.

Step 1: Assembling a Team

An appropriate person for this team does not necessarily have to be an expert in the target area. It could be someone not involved in the core design team but someone the original design team believes can contribute to the development of the flow diagram. Often the participant with the least amount of knowledge about the process provides the most valuable insight into the problem at hand.

Step 2: Identification of the Main Flow Path

A flow diagram typically starts by identifying the main steps of the process as it is currently understood by the team. There are two types of operations commonly used in a flow diagram: a decision step and an action step. Decision steps ask a question and are used to ensure a particular step in the process has been done. Decision steps are represented by a diamond-shaped box. This is shown in Figure 2.3. If the question is answered by a *Yes*, the next step on the main path would be carried out. If the answer is *No*, another pathway must be entered and additional step(s) must be carried out before proceeding to the next step on the main path. This alternate pathway is called a secondary loop. As shown in Figure 2.3, solid or dashed lines with arrows are drawn between each block to show the preferred direction through the process.

Action steps describe the next logical operation in a process and are commonly represented by a rectangular box with either square or rounded corners as shown in Figure 2.4.

There are no set rules for which symbols should or should not be used in a flow diagram. Normally for a flow diagram to be effective it should be kept relatively

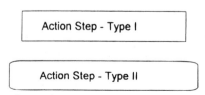

Figure 2.4.   Illustration of an action step.

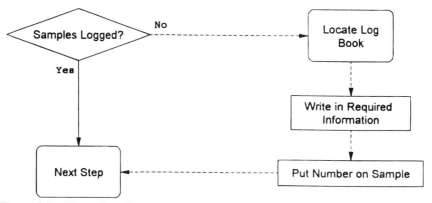

Figure 2.5. Illustration of a secondary loop.

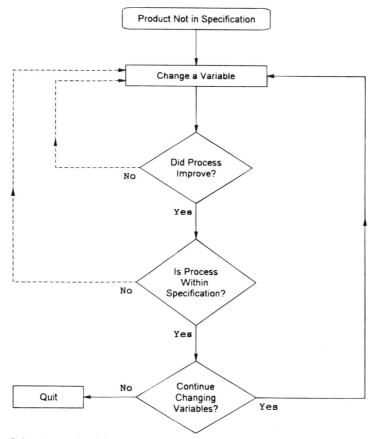

Figure 2.6. A completed flow diagram.

simple. This usually means using only the minimum number of different types of symbols necessary to adequately illustrate the process. In some cases this may not be possible, but simplicity should be the goal.

Step 3: Addition of Secondary Loops

Secondary or decision loops branch out from a decision step located on the main flow path as shown in Figure 2.5. These loops direct or remind a person that an additional step is required before the main flow path can be rejoined and the next step in the main process begun. A secondary loop can also lead to a totally separate path, which is a main flow path itself. For convenience, the secondary loops are typically, but not always, filled in after the main path has been charted. As discussed previously, they are indicated by a line different from what has been used for the main flow path (e.g., broken, dotted, or colored).

Step 4: Check Consistency

Detail used in the flow diagram should be consistent throughout. Unnecessarily complicated flow diagrams defeat their purpose. If a secondary loop is as complicated as the main flow path or contains more than one or two steps, a separate flow diagram should be constructed for the secondary loop. Figure 2.6 is an example of what a completed flow diagram might look like.

3

Data Collection

"There are three kinds of lies: lies, damned lies, and statistics."
Benjamin Disraeli, 1804–1881, English Prime Minister

3.1 Check Sheets

Collection and analysis of data are the cornerstones (and ultimately the solutions) for almost all industrial research projects and manufacturing or marketing opportunities. Each program requires data to develop an appropriate action plan and make decisions. All data collected must be correct and accurate. If the data being gathered are not representative of the process then corrective measures suggested or marketing decisions based on these data are unlikely to be very useful or effective.

An extremely useful and simple tool to gather representative data is the check sheet, which is simply a data collection form used to record the frequency of occurrence of an item or items in a process. As process knowledge increases the check sheet becomes more and more specific. A check sheet must not become overly complicated or difficult to fill out; if it is the required data will not be collected diligently and the whole exercise of data collection will become cumbersome.

3.2 Types of Check Sheets

A check sheet, in its simplest form, consists of a description of the process or product characteristics of interest. When an event occurs (e.g., part failure) or a time is reached, an appropriate representation is made on the sheet. When the time

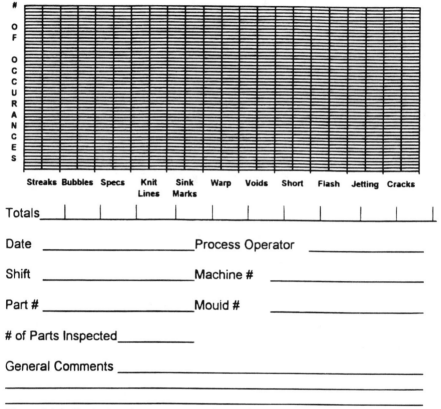

Figure 3.1. Check sheet for a thermoplastics injection molding process.

period decided on for data collection has elapsed, the total of each event would be calculated. An example of a check sheet is illustrated in Figure 3.1.

Check sheets can also be easily adapted to collect data on things such as defect location and distribution of effects. These types of check sheets are illustrated in Figures 3.2 and 3.3, respectively.

3.3 Construction of a Check Sheet

A check sheet can be constructed quite easily. The following is a series of steps that can be followed to complete a check sheet.

Step 1: Selection of the Design Team

As with previous examples, the first step in this process must be to identify and assemble a group of people who are knowledgeable about the process under investi-

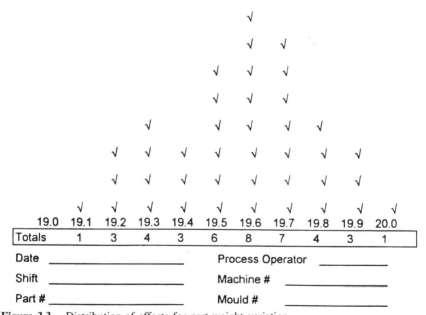

Mark location of defects on diagram

Right side view　　　　　　　　　　Left side view

Back　　　　　　　　　　　　Front　　　Front　　　　　　　　　　Back

Back　　　　　　　　　　　　　　　　　　Front

Top view

Date of Inspection _____　　　　Inspector_____

General Remarks_____

Figure 3.2.　Location of defects.

	19.0	19.1	19.2	19.3	19.4	19.5	19.6	19.7	19.8	19.9	20.0
Totals		1	3	4	3	6	8	7	4	3	1

Date _____　　　Process Operator _____

Shift _____　　　Machine # _____

Part # _____　　Mould # _____

Figure 3.3.　Distribution of effects for part weight variation.

gation. Initially the group should be informed about the project and its function, which is to provide advice. At this first meeting the objectives of the data collection exercise must be established to focus the group's thinking on a specific outcome. For example, if the objective is to reduce the number of scrap batches, the team would design a check sheet capable of easily recording the causes of scrap batches at the time they occurred.

Step 2: Identifying Collection Points

The location for the data to be gathered is another key to making a useful check sheet. The design group should ask itself two questions: Should data be collected in more than one location? and Should identical data collected at different locations be treated separately or combined? Each situation being investigated is bound to have different answers to these questions and should be judged on an individual basis. Consultation of the previously constructed process flow chart is also helpful in deciding appropriate data collection points.

Step 3: Duration of Data Collection

The next important decision is determining the length of time data should be collected. If the time allotted is insufficient the data collected may not be particularly representative of the process. It may also not be possible to establish any sort of process trends. In addition, if it is too long the data collected may not be useful because the opportunity for making changes has been exceeded. Each situation being considered will likely be different. Therefore the design group will have to use its best judgment when selecting time limits for data collection.

Once the overall period for collecting data has been determined, the frequency of data collection within the allotted time must be resolved (i.e., Will the data be taken twice a shift or once a week, month?, etc.). It is usually better to collect too much data at the beginning of the project than to have to repeat the data collection process when it is realized something had been missed. Sufficient data collection time is even more important for processes that have the potential to change with time (e.g., batch processes or fabrication operations).

Step 4: Collecting Data

After designing the check sheet, personnel responsible for collecting the necessary data must be identified. It is advisable to assign more than one individual on each shift to complete this task. The purpose for this is twofold. First, it has the advantage of including more people in the experimental program and allows for some repetition in the data. However, it can also introduce additional experimental error. A check of the results (i.e., comparison of means) should be made to ensure this has not occurred.

Step 5: Data Review

Periodically data should be reviewed by the design team and the people responsible for collection. A continuous review is critical to the collection of appropriate data. It is not always possible to identify appropriate items or categories or even establish a meaningful sampling frequency prior to actually collecting data. If on periodic review the data that have been collected to date are not yielding the desired outcome,

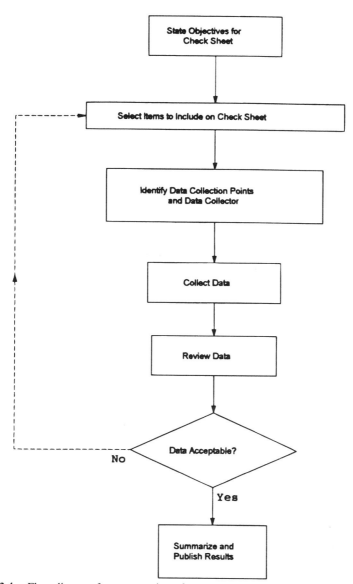

Figure 3.4. Flow diagram for construction of a check sheet.

appropriate changes should be made and the data collection process started over. The data collection process is summarized in the flow diagram given in Figure 3.4.

3.4 Disappearing Income

"I wonder where my money went this month?" For many people, this is not an uncommon question. The obvious result of allowing expenditures to be continually greater than incoming wages is increasing debts. To prevent the accumulation of debts a person might try to find a higher paying job or try to win the lottery, neither of which is especially easy to do. Alternately, what needs to be done is to determine where all one's money goes each month. A check sheet would be very useful for collecting data in this situation. The check sheet for this process would contain categories describing where money could be spent each month as outlined in Table

Table 3.1. List of Monthy Items and
Anticipated Expenditures

Item	Expected Expenditure
Mortgage	500
Car insurance	100
Health insurance	50
Child care	150
Home heating	150
Car care	90
Clothes	120
Cable TV	25
Health care	100
Liquor	80
Retirement fund	110
Vacation fund	75
Life insurance	25
Home insurance	100
City taxes	100
Phone	90
Electricity	80
Misc. house	150
Entertainment	90
Magazines	45
Food	450
Loans	125
Education fund	35
Gifts	60
Total	$2,900

3.1. Included in this table was the anticipated expenditure for each item, such that the total does not exceed the maximum monthly income, of say $2,900.

This is done to be able to compare what was thought was being spent on a particular item to what was actually being spent. Only after there is documentation of how the money is currently being spent will it be possible to assign priorities for spending the monthly income in the future. Once spending priorities have been set, an effective action plan to lower monthly expenditures can be developed.

However, before starting to collect data the time period over which data will be collected must be set. In this example, it is probable that the data collected over a short period of time, say 1 or 2 weeks, would not be representative of actual spending habits. The purchase of one expensive item would tend to distort the sample for that particular category. Data should therefore be collected over a period of several months to be more representative of actual spending habits.

Once a time period for data collection has been established and a suitable list compiled, the actual amounts being spent on each item can be recorded. This is easily done by transferring the itemized list found in Table 3.1 onto a separate piece of paper and keeping it on hand at all times. When a particular item is purchased the amount spent is recorded under the appropriate category. This would continue

Table 3.2. List of Expenditures

Item	Month 1	Month 2	Month 3	Average
Mortgage	500	500	500	500
Life insurance	25	25	25	25
Care insurance	100	100	100	100
Home insurance	100	100	100	100
Health insurance	50	50	50	50
City taxes	100	100	100	100
Child care	150	150	150	150
Phone	115	85	70	90
Home heating	200	150	100	150
Electricity	75	70	95	80
Car care	35	145	90	90
Misc. house	175	125	150	150
Clothes	95	165	90	120
Entertainment	85	65	120	90
Cable TV	25	25	25	25
Magazines	45	45	45	45
Health care	100	125	75	100
Food	450	435	465	450
Liquor	67	123	50	80
Loans	125	125	125	125
Retirement fund	110	110	110	110
Education fund	35	35	35	35
Vacation fund	35	115	60	70
Gifts	35	85	60	60
Totals	$2,832	$3,053	$2,790	$2,892

Table 3.3. Comparison of Estimated Costs versus Actual
Amounts Spent

Item	Estimated	Actual	Deviation
Phone	80	90	+10
Home heating	125	150	+25
Electricity	50	80	+30
Car care	100	90	−10
Misc. house	100	150	+50
Clothes	75	120	+45
Entertainment	50	90	+40
Magazines	10	45	+35
Health care	75	100	+25
Food	400	450	+50
Liquor	75	80	+5
Vacation fund	75	70	−5
Gifts	25	60	+35
Totals	$1,165	$1,525	+$360

Table 3.4. Modified Check Sheet Combining
All Changes

Category	Month 4	Month 5
Living expenses	450	510
Utilities	210	110
Car care	115	75
Travel	45	25
Household	45	50
Savings/loans	135	135
Health	95	80
Other	35	15

until the specified time period for data collection had elapsed. Table 3.2 shows what money was spent for each month and category over a 3-month time period.

It can be seen from these data that the amount spent on numerous items such as the mortgage, insurance, and savings plans does not change month to month. Since expenditures on these items are fixed, there is little value to including them on a check sheet. The amounts spent on these items is critical, however, as they must be deducted from the monthly income to determine the amount available for the variable cost items. Table 3.3 has the fixed expenditure items removed and includes the calculated deviation from estimated amounts for the remaining items.

This is still quite a large check sheet. If practical, it may be possible to group some items together. This could be done to ensure the check sheet is kept simple to use. Categories that could be used in this example would be household products, automotive accessories, health care, entertainment, and treats or recreation rather than listing individual items such as tires, dentist, and movie and baby sitter.

In addition to recording the cost for each item, it may also be helpful to specify how payment was made, cash or charge, especially if there is any indication that method of payment may be partly responsible for high expenditures. Expenditures could also be further categorized into essential and nonessential items.

Combining all the above refinements, a streamlined check sheet would result for the disappearing income example. This new check sheet is presented in Table 3.4.

The information collected can now be used to develop an appropriate plan to regulate spending and ensure there are sufficient funds to cover expenditures.

4

Pareto Charts

"Statistics: I can prove anything by statistics—except the truth."
George Canning, British political leader, 1770–1827

4.1 Pareto Principle

The Pareto principle is based on the work of the nineteenth-century Italian economist Vilfredo Pareto. During his studies at Lucerne University, Pareto observed that certain things were not equitably distributed. Specifically, he found that 80% of the wealth at the time was controlled by only 20% of the people and that 80% of the crime was committed by only 20% of the people (consequently the origin of the 80/20 rule).[1]

Many modern researchers have also found this trend to be true. J. M. Juran was the first to apply this rule to quality-related problems and find the same results. Science in general also adheres to this rule. For instance, the planet Jupiter contains the majority of the orbiting mass of our solar system and the majority of the animal population is made up of only a few species. Distribution of the Earth's human population is also dominated by a handful of countries.

---

[1] Burr, John T., The Tools of Quality Part IV: Pareto Charts. Quality Progress Nov. 1990 pg 59–61.

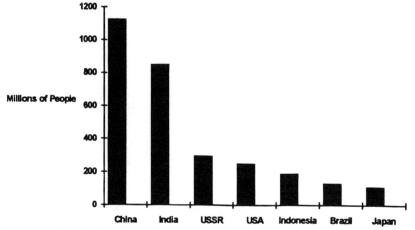

Figure 4.1. Distribution of earth's population (1989). From "The World Atlas" Software Toolworks Inc. © 1989, 1990, 1991.

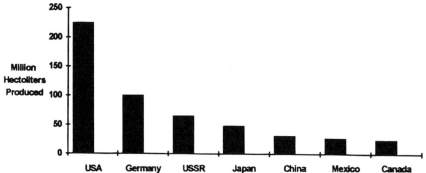

Figure 4.2. Beer production by country (1989). From "The World Atlas." Software Toolworks Inc. © 1989, 1990, 1991.

Even production of a consumer good such as beer adheres to this pattern. These examples are illustrated in Figures 4.1 and 4.2, respectively.

Pareto charts always arrange data in descending order of magnitude from the left of the graph to the right. Therefore, the item that should be investigated first will always be on the left hand side of a Pareto chart. This graphically shows an experimenter which areas require immediate action rather than what is *felt* to be important. It is not uncommon to discover these two can be quite different.

The strength of the Pareto principle does not necessarily lie in exact percentages but in the concept that a very small portion of the many potential contributors to product or process variation actually has a significant effect on the variation. Identification of these few key factors early in a developmental program allows them to be exploited early in the process, thereby maximizing potential system gains.

It is easy to appreciate that time, money, and effort can be wasted if every variable associated with a product or process is rigidly controlled or investigated throughout the experimental program. Sizable economic gains should therefore be possible if the key process/product variables are identified quickly and the nonessential or less significant ones run at their most economic settings.

This, of course, does not imply that the less important variables will be ignored completely. Rather, the effort spent on investigating these variables must be kept in proportion to the time and money allotted for studying a process. Typically it is easier to reduce the biggest problem by 50% than totally eliminate a less significant problem. Elimination of the biggest problem is also more likely to cause a higher return on the investment of time and effort taken.

4.2 Construction of a Pareto Chart

The step-by-step program that follows can be used as a guide in the construction of Pareto charts. A flow diagram for generating Pareto charts is shown in Figure 4.3.

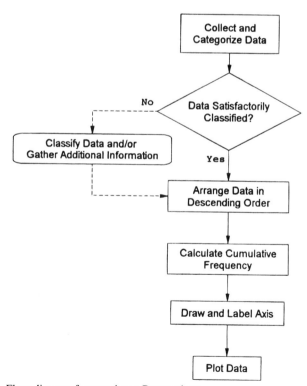

Figure 4.3. Flow diagram for creating a Pareto chart.

Table 4.1. Potential Workplace Injuries

Injury	Quantity
Cuts	8
Leg	12
Back	50
Feet	11
Fumes	5
Hand	5
Head	2
Arm	14
Chest	17
Heat	33

Step 1: Categorize Data

The data to be plotted on a Pareto chart must be categorized into approximately 7 to 10 categories. Using more than 10 categories will unnecessarily complicate the plot. If the data to be plotted are not in a suitably sorted or categorized manner, some reorganization will be required before being able to proceed. In the worst case, it may be necessary to modify or create another check sheet that will collect the required data.

For data to be meaningful in a Pareto chart it cannot be generalized in forms such as the total dollars lost, the number of hours worked in a week, or the number of accidents in a month or year. Consider the data shown in Table 4.1. There is no indication from these data whether any of the injuries were serious, time off work was required, or what the cost was of the treatment needed to repair any damage that occurred. What should be investigated first in this situation? For example, head injuries (two incidents) would be ranked as unimportant if the number of accidents per category was used to determine further action. However, it is not difficult to appreciate that a head injury has the potential to be more serious than a cut finger or broken arm. In this example a system for ranking the injury would need to be developed before a useful Pareto chart could be constructed.

To make a Pareto chart for the total dollars lost per month in a production process, the total dollar value would first have to be subdivided into smaller subtotals such as total dollars lost to down time, operator sickness, low inventories, or poor quality raw materials, as shown in Table 4.2. Only after a dollar figure, or some other representative measurement for each individual heading has been calculated can a useful Pareto chart be constructed. The information in Table 4.2 provides a meaningful data set from which to decide what item should initially be investigated.

Step 2: Formatting

Each Pareto chart requires background information describing the situation being investigated (e.g., title, date, data collection point, data collector, before or after

Table 4.2. Categorized Data
for the Dollars
Lost per Month

Reason	Dollars lost
Customer returns	65,000
Operator experience	50,000
Rework	35,000
Lack of material	21,000
High inventory	17,000
Sickness	14,000
Scrap	9,000
Machine problems	5,000

changes, number of parts inspected, and total number of defects). It may be useful to generate standard forms, similar to that found in Figure 4.4, which can be photocopied and used as required.

It is likely that each situation being investigated will be somewhat different in one aspect or another. Questions such as What time period will be covered by the diagram (i.e., 1 day, 1 week, 1 month, etc.)?, What categories are to be included?, and Who will be using the diagram? need to be considered. When making a blank worksheet allow for flexibility in changing experimental situations.

Step 3: Calculation of Relative and Cumulative Frequency

Pareto charts also frequently incorporate a parameter called the percente relative frequency. This parameter is defined as the percentage of the overall total (i.e., dollars lost, number of defects, calls returned to customers in less than 1 hour, etc.) each category represents. Also of interest is the cumulative frequency, which is simply the cumulative total of the relative frequency for each individual category.

Table 4.3. Calculation of Relative and Cumulative Percent

Reason	Dollars lost	Relative % $\left(\dfrac{\text{dollars lost}}{\text{total}} \times 100 \right)$		Cumulative frequency (%)
Customer returns	65,000	30.1		30.1
Operator inexperience	50,000	23.1	(e.g. $30.1+23.1\Rightarrow$)	53.2
Rework	35,000	16.2	(e.g. $53.2+16.2\Rightarrow$)	69.4
Lack of material	21,000	9.7		79.1
High inventory	17,000	7.9		87.0
Sickness	14,000	6.5		93.5
Scrap	9,000	4.2		97.7
Machine problems	5,000	2.3		100.0
Total	216,000	100		

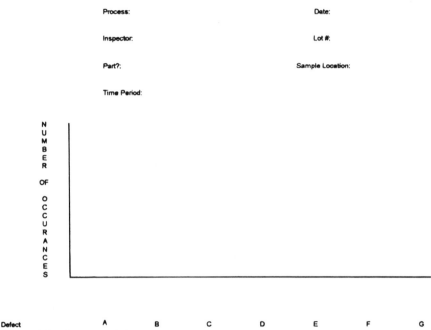

Figure 4.4. Pareto worksheet.

This measure is incorporated into the graph using the right-hand y-axis. The calculations for relative and cumulative relative frequencies are shown in Table 4.3.

Step 4: Plot Data

If the grid for the graph is not already on the blank Pareto sheet then it needs to be drawn at this point. When labeling the grid for the graph, the left-hand vertical

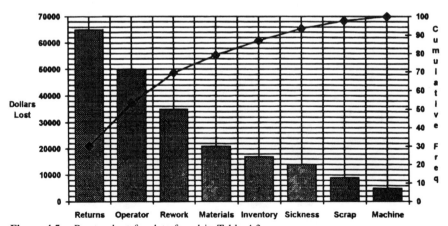

Figure 4.5. Pareto chart for data found in Table 4.3.

or primary y-axis is used for the raw data. The right-hand vertical or secondary y-axis is reserved for the cumulative percent. Once an appropriate scale has been drawn on these axis the collected data can be plotted.

A bar graph is used for the counts or frequency and a solid line is used for the cumulative relative frequency. The largest category is plotted first and is located on the left-hand side of the graph. The next largest category is plotted next and so on until all categories have been drawn on the graph. The only exception to this order is an "other" or "miscellaneous" category, which is always the last one to be plotted. The labels for each category are located on the x-axis. Figure 4.5 would result from plotting the data found in Table 4.3.

CHAPTER

5

Cause and Effect Diagrams

"Quality begins with education and ends with education."

Dr. K. Ishikawa

5.1 Introduction

Cause and effect, C&E, or Ishikawa diagrams were first introduced in the early 1950s by Professor Kaoru Ishikawa. A cause and effect diagram is designed to show how a variable or factor, called the cause, relates to a specific outcome or quality characteristic, called the effect. Professor Ishikawa used these diagrams to teach his students how factors in a process can be sorted and how they relate to one another.

Figure 5.1 shows the skeleton of a cause and effect diagram. Because of their similarity to the skeleton of a fish, these diagrams have also been referred to as fishbone diagrams.

Cause and effect diagrams have many benefits; they

- Increase communication between functions;
- Develop a common understanding about the process;
- Gauge the degree of process knowledge;
- Focus group thinking;
- Can be used for data collection.

Poor communication between functions can sometimes be responsible for many of the misunderstandings that occur. Cause and effect diagrams can play a crucial role in bridging the communication gap between business functions (e.g., manufacturing, sales, marketing, and management) within a company. To generate a cause and

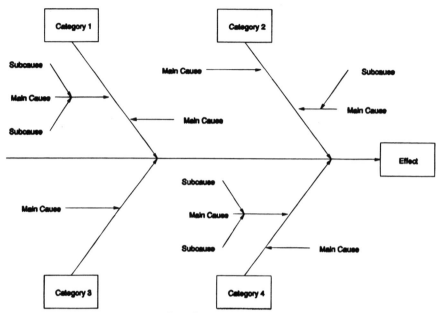

Figure 5.1.   Generic cause and effect diagram.

effect diagram it is generally necessary to assemble a cross-functional group to exchange ideas with one another. This gets people talking and some of the perceived barriers between groups are removed, which ultimately improves communication pathways.

It is unlikely that everyone participating in the development of a cause and effect diagram will have the same level of understanding and knowledge of the process. As the brainstorming session evolves, everyone's understanding of how the different process parameters relate to one another will be enhanced. When completed, cause and effect diagrams provide a neat and simple teaching aid that can be used to train new employees and develop a common knowledge base for veteran employees. A cause and effect diagram can also demonstrate just how well the process is under-stood. A poor quality diagram indicates where process knowledge may be lacking and where additional training may be required.

When developing a cause and effect diagram, the group focuses on only one effect. Potential causes of the effect are generated one at a time. Everyone in the session is given an equal opportunity by the facilitator to suggest causes that influence the effect. As a result, people who are afraid to speak up in a group environment can voice their opinions as effectively as co-workers who regularly speak up. These diagrams can also be used to identify categories of interest for a check sheet.

Another application of cause and effect diagrams is to describe situations that are not related to quality issues or problems. The following list is by no means exhaustive. The only real limitation for cause and effect diagrams is the creativity and motivation of the individuals applying the technique. For example, cause and effect diagrams can easily be developed to

- Identify how a company's sales can be increased;
- Improve a safety program;
- Make a better pancake;
- Establish how to decrease moving traffic violations;
- Improve a person's golf game;
- Improve communication between functions.

5.2 Types of Diagrams

Dr. Ishikawa describes three types of cause and effect diagrams[1]:

- Dispersion analysis;
- Process classification;
- Case enumeration.

These diagrams differ from one another in the way the data are visually presented and how the diagrams are constructed. Preferences for one method over another will depend on the individual or design team carrying out the analysis and the opportunity being explored.

5.2.1 Dispersion Analysis

Diagrams such as illustrated in Figure 5.2 are characteristic of dispersion analysis. Each cause written on the diagram can be used in the following statement: "The effect _____ (e.g., part warpage) could have resulted or have been caused by _____ (e.g., part ejected too soon)." This form of cause and effect allows a direct relationship to be drawn between the opportunity and the potential solutions or causes. It also organizes the causes into distinct categories.

5.2.2 Production Process Classification

Process classification diagrams follow the natural flow of the process of interest from start to finish, much like the flow diagram does. Factors or potential causes that might influence the stated effect should be added at the appropriate site in the diagram. Figure 5.3 represents the process classification approach for the same situation depicted in Figure 5.2. This technique is not limited to the traditional production process and can easily be applied to non-process-related topics as well.

---

[1] Ishikawa, K. "Guide to Quality Control." Asian Productivity Organization Unipub, New York, 1982.

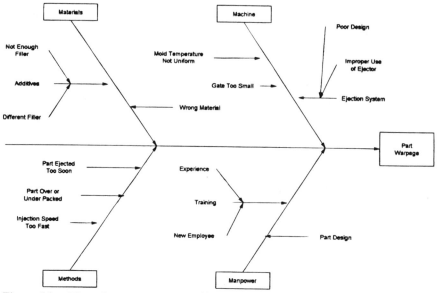

Figure 5.2. Dispersion type cause and effect diagram.

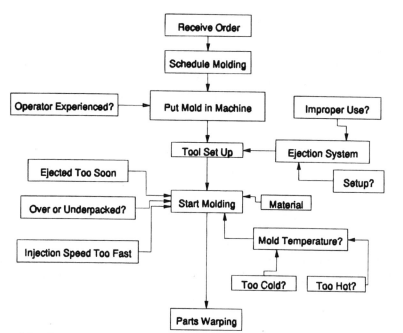

Figure 5.3. Process classification type cause and effect diagram.

A drawback of the process classification approach is that the same cause can be assigned to different stages in the process, which sometimes leads to confusion on the part of the users of the diagrams. In addition, causes attributed to more than one factor can be difficult to depict graphically, making the diagram quite messy.

5.2.3 Case Enumeration

Case enumeration diagrams are generated by having each member of the design team indicate to the facilitator one potential cause for the problem or opportunity being investigated. Each cause is recorded on a sheet of flipchart paper. Once the group can no longer think of any more causes, the causes are organized and written on the traditional fishbone skeleton used in dispersion analysis. This type of cause and effect diagram is sometimes difficult to put together after the session has been completed.

Personal preference, the number of people in the meeting, and time constraints usually dictate which method will be the most effective.

5.3 Construction of Cause and Effect Diagrams

The design team that has been used throughout the preliminary stages of the experimental process is assembled once again. A facilitator must be nominated to direct the group during the brainstorming process. The main purpose of the facilitator is to keep the meeting on track by preventing domination of the session by one individual and encouraging participation by all members. The function of the facilitator is very important in this process and should not be overlooked. A flow chart for the construction of cause and effect diagrams is illustrated in Figure 5.4.

Additional contributors for the development of a cause and effect diagram such as customers, vendors, or operators can be added at this stage to supplement the design team. Customers and vendors are considered at this stage because decisions made at this point often affect them directly. Their input as to their ability to accommodate any proposed changes at this point in the process could prevent an undesirable change from being implemented in the design process.

Several factors contribute to the generation of the best possible cause and effect diagrams. One key factor often overlooked is a good meeting room. The room must be conducive to creative thinking. This means it should have windows, be well lit, have a suitable room temperature, and be insulated from potential interruptions such as phone calls. If no satisfactory meeting room is available in-house then an off-site location could be considered.

Accessories such as felt markers and large sheets of paper for recording and tape for posting the ideas generated by the design team during the brainstorming session should all be readily available. Once a piece of paper has been filled with ideas, it is taped to the walls of the meeting room so everyone can easily read what has been already been written. These points may seem trivial, but in practice they often have a significant impact on the outcome and success of the program.

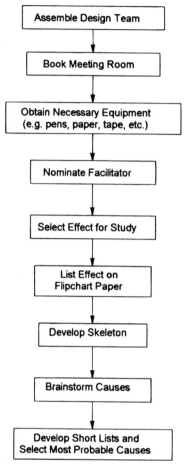

Figure 5.4. Flow chart for constructing a cause and effect diagram.

A precise statement of the problem or opportunity to be used in the cause and effect diagram is the first function to be carried out by the design team. It is the accuracy of this statement that will determine the usefulness of the cause and effect diagram. The objective must then be paraphrased so it can be described in one or two words. Keep in mind there are always two ways of writing an effect. It can either be described as a problem (e.g., "Lousy sales") or as an opportunity (e.g., "What can be done to increase sales?").

Focusing on an opportunity rather than the problem creates a more positive atmosphere in the meeting. Being made to feel a part of the solution rather than part of the problem can be a strong motivation for an individual. It is quite possible that most of the causes generated will be the same regardless of how the effect has been stated. However, some imaginative proposals are likely to be presented because the approach of the meeting has changed from a negative to a positive one.

Once the effect has been adequately described and understood by the design team members, it is displayed in a prominent location in the meeting room. Being displayed in such a manner acts as a constant reminder as to what needs to be accomplished. Typically four to six main categories—materials, machinery, methods, manpower, environment, and measurement—are drawn and linked to the effect to form the main fishbone skeleton as shown in Figure 5.1.

Occasionally, the cause suggested by the design team does not easily fit into one of the standard categories. At this point, the group may want to add another category. Generally speaking, however, 95% of the causes suggested during the brainstorming session can easily fit into one of the four main categories. If possible, the number of additional categories should be minimized as an increase in the number of classifications simply complicates the diagram. Valuable time is wasted discussing under which category to put a cause rather than generating additional causes. When the addition of more categories is unavoidable, no more than six categories is suggested.

Once the main categories have been listed on the cause and effect diagram the subcauses are developed. Regardless of which type of cause and effect diagram is being developed (i.e., dispersion, process classification, case enumeration) each subcause is assigned to the appropriate main category. Brainstorming is extensively used to assist in the generation of probable causes; a guide is given in Table 5.1.

For dispersion and process classification diagrams a line is drawn from the main cause branch first. Then the subcause is written on top of the line as shown in Figure 5.5. A direct relationship between the subcause and the effect can now easily be seen by tracing the line from the cause to the effect (dispersion or process classification) or simply by looking under the appropriate heading in the table (case enumeration).

For case enumeration diagrams the causes can be placed under the appropriate heading as shown in Table 5.2.

When identifying subcauses try not to concentrate on filling up one branch or column at a time. Simply speculate on what could influence the effect. Then as

Table 5.1. A Brainstorming Guide

A clear concise opportunity statement must be developed

Allow the design team time to generate their ideas (this conceptualization time could vary from a few minutes to a few days)

Create as many ideas during the session as possible

No criticism of any idea should be allowed during the brainstorming session

Every member of the design team must be encouraged and given the opportunity to participate

Keep all ideas clearly visible to all members of the team (i.e., mount ideas around the conference room)

When there are no new suggestions from the design team, clarify each point and make sure everyone understands what is meant; if appropriate, combine related ideas

---

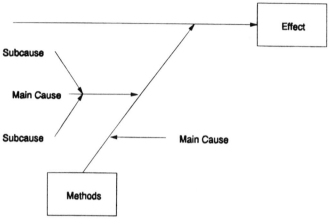

Figure 5.5.   Subcause and main effect relationship.

Table 5.2.   Listing of Subcauses for Case Enumeration
Type Diagrams

Machine	Materials	Method	Manpower
Cause 1	Cause 1	Cause 1	Cause 1
Cause 2	Cause 2	Cause 2	Cause 2
↓	↓	↓	↓
etc.	etc.	etc.	etc.

each new idea comes to mind, it is presented to the facilitator for assignment to the applicable category.

If fewer than 30 causes have been identified consultation with more experienced personnel may be required. It is not unusual, even for the simplest effect or opportunity, to identify over 50 potential causes. A lack of potential causes may also indicate that additional training may be required to improve the group's knowledge of the process. It could also indicate that the effect has not been defined adequately.

Prior to continuing to the next step of selecting the most likely solutions, any causes that are unclear to any members of the design team must be clarified. It is at this point, and not before, that discussions between design group members are allowed. Once the group has finished clarifying and explaining unclear causes, the most likely contributors can be identified.

Reducing the number of causes may involve developing several manageable lists, referred to as *short lists*. The selection process begins by having design team members select a number (5–10 is typical) of causes that they consider to be the most probable contributors to the effect. The 80/20 rule could be used as a guide for the number of causes selected. For example, if 50 causes were identified each

Table 5.3. Potential Contributors to Part Warpage

Cause	Votes
Injection time	✓✓✓✓✓✓✓
Packing	✓✓✓
Part too hot	✓✓✓✓
Shot size	✓✓✓
Gate size	✓✓✓✓
Injection pressure	✓
Mold temperature	✓✓✓✓✓
Melt temperature	✓✓✓✓✓✓✓
Thickness	✓✓
Part ejection	✓✓✓
Mold dimensions	✓✓✓✓

person would be asked to write down 10 causes. If 40 causes were identified each person would choose only 8 causes, and so on.

Each selection is written on a piece of paper or file card and given to the facilitator. Each choice is then copied onto flipchart paper. If a cause happens to be chosen by more than one person, a check mark is placed beside that cause to indicate that it has been selected more than once. To illustrate this, consider the list, shown in Table 5.3, generated during a brainstorming session looking at the potential causes of warpage in thermoplastic parts.

The design team should now prioritize and scrutinize the short list to select factors for the designed experiment. At this point it is necessary to reach a consensus as to which factors should be studied. Costs, practicality, and the likelihood of implementing the changes can be used as criteria to evaluate the desirability of using a particular factor in the proposed experiment. The chosen factors are then summarized in a *Trial List Table* as shown in Table 5.4.

Cause and effect diagrams need not be generated within the structured confines of a brainstorming session. A much less structured approach can be used. Remember,

Table 5.4. Trial List

Cause	Votes
Injection time	✓✓✓✓✓✓✓
Melt temperature	✓✓✓✓✓✓✓
Mold temperature	✓✓✓✓✓
Gate size	✓✓✓✓
Part too hot	✓✓✓✓
Mold dimensions	✓✓✓✓
Part ejection	✓✓✓

however, that this method should be used only as an occasional alternative to complement normal brainstorming techniques.

A bare fishbone diagram containing all four main categories (i.e., manpower, materials, machine, and methods) and the effect of interest is displayed in an area highly visible to all employees. When someone thinks of a factor that might influence the effect they can write it down on the diagram.

Typically, the diagram would be left up for approximately 1 to 2 weeks,[2] giving all employees ample time to express their thoughts. After the specified time has elapsed, the diagram would be removed. The design team would then review the diagram and devise the appropriate strategy to address the effect.

---

[2] Sarazer, J. Stephen., The Tools of Quality. Part II: Cause and Effect Diagrams, Quality Progress July 1990, pg 59–62.

CHAPTER

6

Basic Statistics

"Do you want the truth or do you want statistics."

Mark Twain

6.1 Introduction

Quotes, like the one above, and at the beginning of each chapter demonstrate the cynicism people have had with statistics throughout history. Simply mention the word statistics to many scientists and engineers and it summons up the most unfavorable response. According to Webster's "Ninth New Collegiate Dictionary" statistics are "a branch of mathematics dealing with the collection, analysis, interpretation, and presentation of masses of data." So why is there such mistrust or dislike for this subject?

This skepticism toward statistics or statisticians may have some of its founding in the way statistics have been used. For example, it is not uncommon to hear an advertiser say: "Nine out of ten people are satisfied using our product and would use it again." These statements are, undoubtedly, the absolute truth from the advertisers' point of view. However, this does not guarantee that the group of people who tested the product are in any way representative of the typical consumer. Regardless of the mistrust or unease with statistics, they are a part of our daily lives and most of us use them on a regular basis.

The morning weather report quotes the average daily temperature, What sports fan doesn't go directly to the sports page each morning to find out their favorite baseball player's batting average or what a pitcher's save percentage may be. There are those who like to bet on sports and hurriedly look for the column where the odds are available on whether a team will win or lose a particular sporting event or by how much they will beat their opponent. If a consumer is in the market for a particular article, such as a car, consumer reports are reviewed to determine which brand on average has a history of reliability, safety, and resale value. Financial newsletters describe the average percentage return on a particular stock or mutual fund over a given period of time to entice stock purchases from potential investors.

A rudimentary understanding of some basic statistical measures such as *mean, median, variance, standard deviation, samples, populations,* and *distributions* can be extremely helpful in understanding the world of statistics. Anybody who carries out experiments in their day-to-day job will need data to make decisions and cannot avoid the use of some of these basic measures.

6.2 Descriptive Statistics

Descriptive statistics are designed to summarize important features of a given data set. People begin to have problems with descriptive statistics when they attempt to draw conclusions beyond the scope of the measure being used. For example, the listing of how a mutual fund has done in the recent past is one way mutual fund companies describe their product. However, this statistical measure of past performance (i.e., average yield) does not in any way guarantee a similar return in the future. The average yield simply describes or summarizes the key aspects of the fund over the time interval the data were collected. It is not possible to make any additional statements about future performance without additional information.

Consider another example. If the distances traveled between each of the last 10 fill-ups of the family car were 235, 275, 325, 245, 267, 287, 297, 241, 235, and 223 miles, it would be quite correct to say that 50% of these values were below 250 miles and 50% were greater. It would be equally accurate to state that the average mileage during the last 10 fill-ups was 263 miles. However, it would go significantly beyond the scope of the statistic measure used to say that the average distance that could be traveled between fill-ups for this particular make and model of vehicle would be 263 miles. There is no knowledge of the types of roads traveled, what kinds of tires were used, the skill level of the driver, and so on, which are all quite likely to affect the mileage being obtained.

6.3 Samples and Populations

According to our definition, statistics look at masses of data. If the data set contains all conceivable observations pertaining to a specific parameter then it would be called a *population*. If the data were only a subset of these observations then it would be considered a *sample* of the overall population. One of the key uses of statistics is to make an inference about a population based on the information provided by a sample.

Imagine that the decision to buy 100 baseball bats from a particular bat supplier depends on their adherence to a weight specification. For instance, the average weight of 10 bats could be used as an indicator of the actual weight of each of the 100 bats. The 10 bats would be called a sample of the population of 100 bats. Now if the 100 bats were only the first of many more orders of bats then the weight of the first 100 bats becomes a sample of the greater population of bats yet to be supplied.

6.4 Average

One of the ways statistics are used to characterize a data set or a sample is to calculate a few meaningful descriptors. One measure that should be familiar to most people is the *average*. An average is used as an indicator of where most of the data in a sample are clustered. It is also called the mathematical center of the data. For a population the word used to describe the center of the data is the mean. The popularity of the sample average arises because it can be calculated for every data set, only one average exists for each sample, and all data in the sample are used in the calculation. If the sample size is sufficiently large it is also generally a good predictor of the population mean.

Mathematically defined, the average is the sum of all the data in the sample, divided by the total number of data points in the data set. Algebraically, the average can be represented by Eq. (6.1).

$$average = \frac{x_1 + x_2 + x_3 + \cdots + x_n}{n}$$

(6.1)

Each x in Eq. (6.1) represents a single data point in the sample. n denotes the total number of data points in the sample. Therefore, x_1 is the first data point in the sample, x_2 is the second, x_3 is the third, and so on until x_n is reached, which naturally represents the last data point in the sample.

By using the Greek alphabet, formulas can be presented in a more compact fashion. The Greek letter sigma, Σ, is used to denote the sum of individual data points and $\bar{X}$ represents the sample average. Incorporating these Greek symbols into Eq. (6.1), Eq. (6.2) would be obtained.

$$\bar{X} = \frac{\Sigma x_n}{n}$$

(6.2)

6.5 Mean

A convention has been adopted for denoting the differences between sample measures and population measures. Generally, Greek letters are used to describe a population and Latin or conventional lower case letters are used when describing samples. Therefore the equation for a population mean would be similar to the sample average but a Greek symbol would replace the Latin letter as is shown in Eq. (6.3).

$$\mu = \frac{\Sigma \, x_n}{N} \qquad (6.3)$$

In this equation the Greek letter μ is used to represent the population mean and N is used for the total number of data points in the population. Calculation of an average is demonstrated in the following example. Consider the data found in Table 6.1, which represent a sampling of the time it has taken a person to get to work.

The mean of these data would be

$$\overline{X} = \frac{\Sigma \, x_n}{n}$$

$$= \frac{19 + 21 + 18 + 38 + 25 + 23 + 24 + 29 + 19 + 22 + 26}{11}$$

$$= \frac{264}{11} = 24$$

The calculation of an average can suffer from one important weakness. If the data are skewed to one side of the average, the sum obtained is unlikely to be representative of the true value of the center point of the data, especially in samples that have relatively few data points (i.e., less than 25). It is always advisable to work with larger groups of data, 25 or more, to minimize the impact of large spreads between data points. This phenomenon is graphically illustrated Figure 6.1.

6.6 Median

Consider the price of four houses for sale in one particular residential area. The prices for these houses are $164,000, $115,000, $50,000, and $60,000. An average

Table 6.1. Time Required to Drive to Work

Day	Time (min)
1	19
2	21
3	18
4	38
5	25
6	23
7	24
8	29
9	19
10	22
11	26

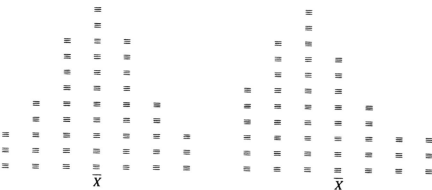

Figure 6.1. Data with the same mean but different distributions for the data within the sample.

sale price for these houses would be $97,250, which is not particularly representative of any one house for sale in this area. It is sometimes helpful to use the median or the middle value of the data set instead of the average.

To calculate a median the data must be arranged in either descending or ascending order. The median would then be defined as the value occupying the middle position in the sample data. In other words, it is the point at which 50% of the values are greater than this number and 50% are lower. When there is an even number of data points, the average of the two middle values of the sample set is used as the median. With an odd number of data points the median will be the actual value of the middle data point.

Equation (6.4) gives the formula for calculating the median.

$$\overline{\overline{X}} = \frac{n + 1}{2}$$

(6.4)

where $\overline{\overline{X}}$ = the median location in the data set
n = the number of data points in the sample

Using the information from the housing prices example above, the location of the median value of the data would be found as follows:

$$\overline{\overline{X}} = \frac{4 + 1}{2}$$

$$= \frac{5}{2} = 2.5$$

= average between the second and third number in the data set

− $87,500

If the data set contained an odd number of data points, say 25, the median location would be calculated as follows:

$$\overline{\overline{X}} = \frac{25 + 1}{2}$$

$$= \frac{26}{2} = 13$$

Therefore the median would be represented by the thirteenth data point in the sample.

6.7 Variation

Imagine that an experiment had been conducted to determine how different alcohol consumption rates affected the percent alcohol in a person's bloodstream. In one sample of volunteers, one alcoholic beverage was consumed by each person in the group every hour for 12 hours. A second group of volunteers consumed two beverages in each of the first 6 hours of the test and then nothing for the remaining 6 hours. A blood sample was taken from each subject in the study every 2 hours and the percent alcohol in the blood was determined. Results from this experiment are summarized in Table 6.2. Each value listed is an average of 10 subjects. This calculation is not shown in this example.

Considering only the average of the data in this example, the different alcohol consumption rates would appear to have had similar effects on the test subjects. However, it is not hard to imagine that the different doses of alcohol could have

Table 6.2. Alcohol (%) in the Bloodstream

Hour	Group 1[a] (12 × 1 drink/hr)	Group 2[b] (6 × 2 drinks/hr)
2	0.05	0.05
4	0.06	0.10
6	0.08	0.14
8	0.08	0.11
10	0.10	0.06
12	0.11	0.02

[a]Group 1 average:

$$\overline{X} = \frac{0.05 + 0.06 + 0.08 + 0.08 + 0.10 + 0.11}{6}$$

$$= \frac{0.48}{6} = 0.08$$

[b]Group 2 average:

$$\overline{X} = \frac{0.05 + 0.10 + 0.14 + 0.11 + 0.06 + 0.02}{6}$$

$$= \frac{0.48}{6} = 0.08$$

had radically different effects on the various test subjects. Visual inspection of the data shows that the individual data points in Group 1 are different from Group 2. This suggests that the average by itself may not be adequate to completely describe a data set, but a measure that describes the breadth or variation present in the data is also required. In other words, before a group of data can be accurately described, an idea of how similar the individual data points within a data set are to one another is needed.

6.8 Standard Deviation and Variances

One possible way to describe the spread between raw data could involve finding the average of the deviation from the overall sample average. To do this, the average of the data set would be subtracted from each individual data point in the sample. These values would then be averaged together to get the average deviation from the sample average.

However, when this calculation is carried out an obvious drawback becomes noticeable. Subtracting the mean from each individual data point results in both positive and negative numbers, as some of the data are naturally greater than and some data are less than the mean. When these values are added together the resultant average sums to zero, which, of course, is not particularly helpful.

How can this problem be overcome? A common and useful mathematical technique used to eliminate negative numbers is to square the number, or in other words, multiply the number by itself. When two negative or positive numbers are multiplied together a positive response will always be obtained. Manipulation of the data in this fashion is allowed only if all the data in the sample are treated in an identical manner.

Using the data from the alcohol experiment, this approach of squaring the deviation from the average is demonstrated. First the average is subtracted from each data point then multiplied by itself or squared. Carrying this procedure out on each data point in Group 1, the following values result:

$$\Delta x_1^2 = (0.05 - 0.08)^2 = 0.0009 \qquad \Delta x_2^2 = (0.06 - 0.08)^2 = 0.0004$$

$$\Delta x_3^2 = (0.08 - 0.08)^2 = 0.0000 \qquad \Delta x_{4^2} = (0.08 - 0.08)^2 = 0.0000$$

$$\Delta x_5^2 = (0.10 - 0.08)^2 = 0.0004 \qquad \Delta x_6^2 = (0.11 - 0.08)^2 = 0.0009$$

Each value is now positive and the mean of the values can now be found. This equation is also commonly referred to as a sample variance and is represented by the symbol s^2 in classical statistics and by V_x in Taguchi terminology. The mean sum of squares or variance calculation is summarized by Eqs. (6.5a) and (6.5b).

$$s^2 = \frac{\sum\limits_{i=1}^{n} (x_i - \overline{X})^2}{n - 1} \tag{6.5a}$$

$$V_x = \frac{\sum\limits_{i=1}^{n} (x_i - \overline{X})^2}{n - 1} \tag{6.5b}$$

Using this formula to calculate the variance of the blood alcohol experiment, the variance for Group 1 data would be

$$S^2 = \frac{0.0009 + 0.0004 + 0.0000 + 0.0000 + 0.0004 + 0.0009}{6 - 1}$$

$$= \frac{0.0026}{5} = 0.000519$$

A value of 0.00196 would be obtained if this same calculation was carried out for Group 2 data. This value of course is much greater than the Group 1 variation. Calculation of this measure of variation enables a distinction to be made between the two groups of data. Statistically speaking, the mean squared deviation (MSD_x) or mean sum of squares (MSS_x) has just been calculated.

Variance is a very useful tool in determining how the data are dispersed about the mean, however, the data are no longer in the original units of measurement because they have been squared. It is necessary at times to convert the variance back into its original units. This can be accomplished by taking the square root of the number. When the square root is taken of a number that has been squared, the number or symbol along with the original units will remain. When the square root of both sides of Eq. (6.5) is taken, Eq. 6.6 results. This alternate measure of variation is now referred to as the *sample standard deviation:*

$$\sqrt{s^2} = \pm \sqrt{\frac{\sum\limits_{i=1}^{n} (x_i - \overline{X})^2}{n - 1}}$$

$$s = \pm \sqrt{\frac{\sum\limits_{i=1}^{n} (x_i - \overline{X})^2}{n - 1}} \tag{6.6}$$

A similar formula for the population standard deviation can also be written by once again using the Greek alphabet. σ replaces s, μ is exchanged for $\overline{X}$, and N replaces $n - 1$. The equation for a sample standard deviation becomes Eq. (6.7) for a population.

$$\sigma = \pm \sqrt{\frac{\sum\limits_{i=1}^{n} (x_i - \mu)^2}{N}} \tag{6.7}$$

6.8.1 Calculation of the Standard Deviation

Calculation of the standard deviation by hand can be quite simple if the following steps are completed:

- Calculate the average for the data set;
- Calculate the deviation from the average for each data point;
- Square each deviation from the average;
- Sum the squared deviations together;
- Divide the total sum of the squared deviations by $n - 1$;
- Take the square root of the average of the squared deviations.

Consider a couple who are planning to retire and are considering two cities in which to live. They have been provided an average monthly as well as the yearly average temperature as shown in Table 6.3. Using only this limited information, let us provide the couple with one additional piece of information, the standard deviation, to assist them in their selection process.

The calculations for the standard deviation are shown in Table 6.4. As a check of the calculations, if the sum of all deviations from the average does not total 0, then a mathematical error has been made and would need to be corrected before proceeding with additional calculations.

Calculating the total squared deviation from the average, the following results:

$$\text{City}_1 = \frac{4300}{12 - 1} = 390.9$$

$$\text{City}_2 = \frac{334}{12 - 1} = 30.4$$

Table 6.3. Average Monthly
Temperatures
for Two Cities

City 1 (°F)	City 2 (°F)
50	69
60	70
70	71
75	74
80	76
90	79
105	83
110	85
80	78
75	76
55	71
50	68
Ave = 75	Ave = 75

Table 6.4. Squared Deviations from the Average for Retirement Example

	City$_1$			City$_2$	
x_i	$(x_i - \bar{X})$	$(x_i - \bar{X})^2$	x_i	$(x_i - \bar{X})$	$(x_i - \bar{X})^2$
50	−25	625	69	−6	36
60	−15	225	70	−5	25
70	−5	25	71	−4	16
75	0	0	74	−1	1
80	+5	25	76	1	1
90	+15	225	79	+4	16
105	+30	900	83	+8	64
110	+35	1225	85	+10	100
80	+5	25	78	+3	9
75	0	0	76	1	1
55	−20	400	71	−4	16
50	−25	625	68	−7	49
Total	0	4300		0	334

To complete the standard deviation calculation, the square root of the average squared deviation from the mean must be taken.

$$\text{City}_1 = \pm\sqrt{390.9} = \pm 19.8$$
$$\text{City}_2 = \pm\sqrt{30.4} = \pm 5.5$$

If the standard deviation is large it indicates the data are broadly dispersed around the average. Conversely, a low standard deviation indicates the data are closely grouped around the average. In this example, City$_1$ has a standard deviation that is 3.5 times as great as that of City$_2$. This means that the temperatures City$_1$ experiences are not very close to the average and may not be as desirable a place to live for the retiring couple as City$_2$ where the temperature is fairly constant year round.

6.8.2 Alternate Standard Deviation Equation

A standard deviation can also be calculated using Eq. (6.8).

$$s = \pm\sqrt{\frac{\sum_{i=1}^{n} (x_i)^2 - [(\Sigma\, x_i)^2/n]}{n - 1}} \tag{6.8}$$

Use of this formula will yield a result identical to the previous standard deviation formula. A numerical check on this formula is made by using the data for City$_1$ found in the previous example. These calculations are summarized in Table 6.5.

Table 6.5. Verification That Standard Deviation Eq. (6.8) Is Equivalent to Eq. (6.6)

x_i	x_i^2	$\dfrac{(\sum x_i)^2}{n}$	$\sqrt{\dfrac{\sum (x_i^2) - \left[\dfrac{(\sum x_i)^2}{n}\right]}{n-1}}$
50	2,500		
60	3,600		
70	4,900		
75	5,625		
80	6,400		
90	8,100	$\dfrac{(900)^2}{12}$	$\sqrt{\dfrac{71,800 - 67,500}{12 - 1}}$
105	11,025		
110	12,100	$= 67,500$	$\sqrt{\dfrac{4,300}{11}}$
80	6,400		
75	5,625		± 19.8
55	3,025		
50	2,500		
Total 900	71,800		

6.9 Normal Distribution

It has just been shown mathematically how to describe the way data are dispersed about the average. Experimental observations vary for any number of reasons, such as humidity, natural laws, and experimental error. In general the probability these variations will occur are essentially symmetrical about the mean and small deviations are far more frequent than large ones. In fact, about 96% of all deviations from the mean occur within two standard deviations of the mean. A common distribution encountered in industry is illustrated in Figure 6.2.

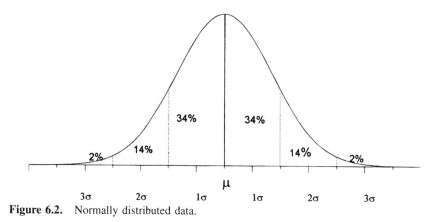

Figure 6.2. Normally distributed data.

Distributions, as shown in Figure 6.2, are referred to as normal distributions and will be the most commonly encountered type of distribution. This will be regardless of whether you are comparing the physical attributes (e.g., hair color or shoe sizes) of a given population, the time it takes to travel to work, or the batting averages of your favorite baseball players. Strictly speaking, however, few populations are *exactly* normal, but many approach normality closely enough to allow the application of the rules of a normal distribution. Assuming that a population follows a normal distribution greatly simplifies the statistical approach that can be taken.

A distribution is completely described[1] by the mean and standard deviation values and each combination of these two measures generates a unique distribution. The center of the distribution is set by the mean and the shape (e.g., narrow or wide) is specified by the standard deviation. Figures 6.3 illustrates how the mean and standard deviation interact with one another.

As discussed previously, a very large portion of the data is grouped closely around the mean. A theoretical characteristic of the normal distribution is that 68% of all observations can be found within 1 standard deviation of the mean, 95.7% will be found within 2 standard deviations, and 99.7% of the data will be within 3 standard deviations of the mean. This characteristic of a normal distribution is illustrated in Figure 6.4

6.10 Hypothesis Formulation

Comparisons of one type or another are common in industry: evaluation of alternate materials aimed at increasing the yield of a reaction, testing of new equipment aimed at increasing production rates, and determining whether additive A works better than B are typical examples. In actuality, the experimenter is trying to prove (or disprove) that the "alternate" treatment, process, or additive being evaluated does not cause a result different than what is presently being done. In statistical terms this is stating the *null* (H_0) hypothesis.

A null hypothesis states that any differences in observed output between the two process treatments are statistically insignificant and therefore cannot be attributed to the change made in the *process* but is more likely the result of random chance or experimental error. Algebraically, the null hypothesis can be stated by Eq. (6.9).

$$H_0 = \mu_1 = \mu_2 \tag{6.9}$$

where μ_1 represents the mean of the first data set and μ_2 represents the mean of the second data set.

---

[1] A normal distribution is defined by the following formula, where the only parameters that are not constant are the mean, μ, and standard deviation, σ.

$$f(x) = \frac{1}{\sigma\sqrt{2\pi}}\, e^{-1/2[(x-\mu)/\sigma]^2}$$

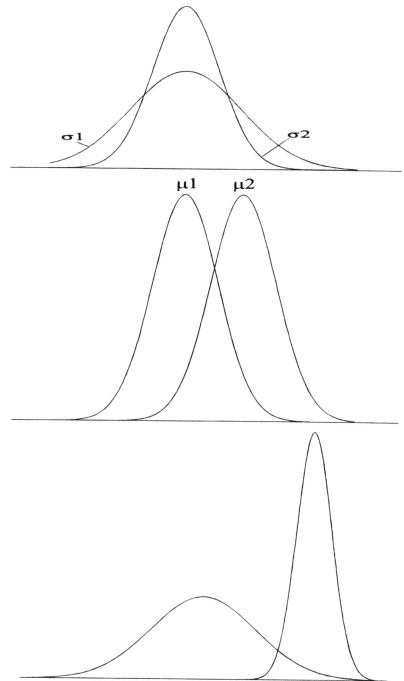

Figure 6.3. Distributions with (a) the same mean but different standard deviations and (b) the same standard deviation but different means. (c) Normal distributions with different means and standard deviations.

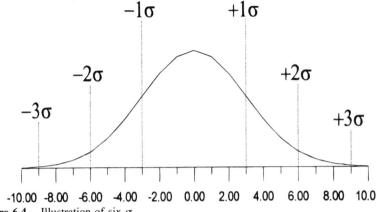

Figure 6.4.   Illustration of six σ.

The converse to the *null* hypothesis is the *alternate* hypothesis. The alternate hypothesis states that there is sufficient evidence to conclude that a statistically significant difference does exist between the two materials, products, or processes being investigated. In other words, the observed output differences cannot be entirely attributed to random chance and are likely a result of the change made to the *process*. Algebraically the alternative hypothesis can be stated by Eq. (6.10).

$$H_a = \mu_1 \neq \mu_2 \tag{6.10}$$

where μ_1 once again represents the mean of the first data set and μ_2 represents the mean of the second data set.

These hypotheses form the basis for decision making. If, after performing the experiments, the difference in the response as a result of the two different treatments is too large to be attributed to random error the null hypothesis would be rejected in favor of the alternate hypothesis. If, on the other hand, the difference was small and could reasonably be attributed to random chance then the null hypothesis would be accepted.

The next logical question resulting from this discussion is how to determine whether a difference is statistically small or large enough to either accept or reject the null hypothesis. This begins with the assessment of risk and setting confidence levels.

6.11 Error and Risk

There is always a possibility when accepting the null or alternate hypothesis that the conclusion obtained is in error. Two types of error, *Type I* or *Type II*, can be made. A *Type I* error is committed if a "treatment" is considered to have a significant effect on the process when it really does not. A *Type II* error results when a "treatment" is considered not to be significant when it really is a major contributor to output differences. This concept is illustrated below.

	Decision Point	
Truth	Accept H_0	Don't Accept H_0
H_0	H_0 is correct and not rejected; there is no problem	H_0 is correct but is rejected; there is a problem (Type I error)
H_a	H_a is correct but is rejected; there is a problem (Type II error)	H_a is correct and not rejected; there is no problem

The probability of committing a Type I error is established by choosing an appropriate alpha risk (α-risk). Choosing an α-risk is very important as this will define the minimum level of protection against accepting the null hypothesis when in fact the alternate hypothesis is true. An α-risk value is chosen prior to collecting data and in most instances is an arbitrary choice. Generally, for noncritical applications, an α-risk of either 0.1 or 0.05 is considered acceptable. However, for applications demanding a greater degree of precision and/or accuracy or if it is a safety-related investigation, an α-risk value of 0.01 or greater should be considered.

A confidence level for the experiment is then generated by using the formula (1 − α) and is typically expressed as a percent. So if a 0.05 α-risk was chosen, the confidence level would be 95%. This means that 95 times out of 100 (commonly expressed as 19 times out of 20 in the media) the decision made will be the correct one. An F test can then be used to determine if a hypothesis is statistically significant.

6.12 F Statistic

To carry out a test to determine the statistical significance of a result, an F ratio for the experimental data must be calculated. An F ratio is the ratio of the measured variances for the different treatments in the experiment as shown in Eq. (6.11). The largest variance is always found in the numerator of the equation.

$$F_{(confidence\ level)} = \frac{s^2_{Large}}{s^2_{Small}} \tag{6.11}$$

For example, if the confidence level desired was 95% and $s^2_1 = 5.22$ and $s^2_2 = 3.13$, then the F ratio for this combination would be

$$F_{95\%} = \frac{5.22}{3.13} = 1.67$$

Next, a theoretical F ratio must be obtained. The applicable theoretical F ratio can be found by

• Going to an F Table (Appendix D) that corresponds to the chosen confidence level;

- Following along the top (horizontally) of the table until the column corresponding to the value that is one less than the total number of specimens tested is found for the sample with the largest variance;
- Continuing down (vertically) that column until the row corresponding to the value that is one less than the number of specimens tested for the sample with the smallest variance is found.

If, for example, the sample with the largest variance has three specimens and the one with the smallest variance has five specimens, what would the theoretical F ratio be assuming an α-risk of 0.05? Following the above procedure column 2 would be required for the large variance sample and row 4 for the parameter corresponding to the factor with the smaller variance. In this example, the F ratio would be 6.94. This scenario is shown graphically in Figure 6.5.

Once the theoretical F ratio has been found it is compared to the calculated F ratio. If the calculated F ratio from the experimental data is smaller than the one found in the table, then the null hypothesis would have to be accepted and the alternate hypothesis rejected. If the converse was true and the experimental F ratio was larger, a statistically significant relationship is indicated and the null hypothesis should be rejected and the alternate hypothesis accepted.

Consider the following example of a balloon-making operation. A new latex supplier has approached the balloon maker and claims that the use of his compound will result in a stronger balloon. The balloon manufacturer decides to try the new latex formulation. An experiment is set up in which a quantity of balloons will be made using the old material and an equivalent number of balloons manufactured using the new material. The tensile strength of both sets of balloons is then tested.

Stating the null hypothesis for this experiment: "The new latex formulation will perform no differently than the presently used formulation and any observed differences in balloon tensile strength can be attributed to random error or chance."

factor with the largest variance $\Rightarrow$

factor with the smallest variance ⇓	1	2	3	4	5	6
1	161	200	216	225	230	234
2	18.5	19.0	19.2	19.2	19.3	19.3
3	10.1	9.55	9.28	9.12	9.01	8.94
4	7.71	6.94	6.59	6.39	6.26	6.16
5	6.61	5.79	5.41	5.19	5.05	4.95

Figure 6.5. Determination of theoretical F ratios.

The alternate hypothesis would state: "The measured differences in the balloon strengths are statistically significant and the differences in balloon strength cannot be explained by random chance or error." A confidence level of 95% is considered acceptable for this experiment.

Where

$$H_0: \mu_1 = \mu_2$$
$$H_a: \mu_1 \neq \mu_2 \qquad \alpha = 0.05$$

The standard material resulted in an average tensile strength of 5236 with a standard deviation of 150 psi on 10 balloons tested. An average tensile strength of 6000 with a standard deviation of 25 psi was measured for the new material, also on 10 balloons tested. Should the null hypothesis be accepted or rejected?

Calculating the F ratio for this experiment at the 95% confidence level:

$$F = \frac{150}{25} = 6.00$$

The theoretical F ratio obtained from Table 1 in Appendix D would be 3.18 since 10 balloons were tested for both products. This suggests, because the experimental F ratio is greater than the theoretical one, that the new formulation produces a balloon that is different from the standard formulation. Therefore the null hypothesis should be rejected in favor of the alternate hypothesis. The larger the difference between a calculated and theoretical F ratio the more confidence there is that the two samples are different.

7

Analysis of Variance

Statistics should be used to diagnose the patient rather than to cure the ailment.

7.1 One-Way Analysis of Variance

In most industrial experiments it will be necessary to compare more than two sets of averages. A commonly used method to compare results from more than two sets of averages is the analysis of variance or ANOVA technique. In the ANOVA calculation, the magnitude of the change in a measured response[1] is compared to the magnitude of the calculated experimental error rather than being directly compared to a sample mean. The following example on fuel efficiency demonstrates how a one-way ANOVA calculation is performed.

7.2 Changing Fuel Efficiency

Fuel efficiency or gas mileage of a motor vehicle tends to fluctuate. Does this variation result from something beyond the driver's control or could it be directly attributable to something that may be controllable? To answer this question an

---

[1] A measured response is also referred to as a response variable, dependent variable, or quality characteristic.

experiment must be conducted. A cause and effect diagram, listing the factors that might contribute to variations in fuel efficiency, is constructed to begin this investigation. Figure 7.1 is an illustration of the cause and effect diagram generated.

After studying this diagram it was reasoned that the gasoline retailer could be a strong contributor to the deviation in fuel economy. The null and alternate hypothesis for this experiment would be as follows:

- Null hypothesis: No substantial differences exist between the gasolines obtained from the different retailers

    H_0: $\mu_1 = \mu_2$

- Alternate hypothesis: A significant difference exists between the gasolines obtained from the different retailers

    H_a: $\mu_1 \neq \mu_2$

If a particular retailer supplied superior gasoline it should be quite easy to consistently use that one retailer and reap the benefits of improved fuel efficiency. This of course assumes that the fuel delivered to the retailer will be consistent and that batch-to-batch variations in gasoline quality will be minimal. Collection and analysis of the data for this experiment were kept relatively simple by looking at only three different gasoline retailers and only one grade of gasoline. Information was collected on the distance traveled between fill-ups, the grade of gasoline used, the quantity of gasoline purchased, the retailer, as well as general driving conditions. In addition, the gas mileage was calculated after each gas purchase and recorded on the check sheet as

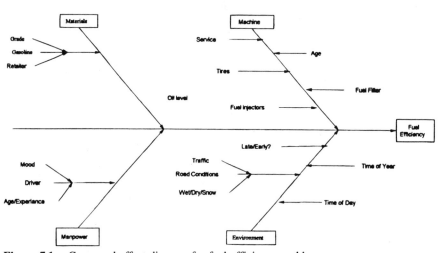

Figure 7.1. Cause and effect diagram for fuel efficiency problem.

Table 7.1. Sample of the Check Sheet Used for Data Collection in the Gasoline Experiment

Date	Leave Home or Work	Arrive Work or Home	Travel Time	Weather	Delays?	Route	Service Station	Odometer Reading	Fuel Needed This Fill	Mileage (liters/ 100 km)
Aug 20	8:08	8:27	19	Dry	None	2	—	—	—	—
20	17:00	17:30	30	Dry	Slow	3	A	1883.4	54.45	8.65
21	7:54	8:13	19	Dry	None	2	—	—	—	—
21	17:24	17:55	31	Dry	None	3	—	—	—	—
24	7:40	8:05	25	Dry	None	3	—	—	—	—

well. Data were collected for a 1-year period. An example of the check sheet used to collect data for this experiment is shown in Table 7.1.

In Table 7.2 a summary of the results from a random selection of fill-ups for the test vehicle obtained throughout the year is presented.

7.3 Types of Variation

Three types of variation can be identified within a data set. There is the total variation in the individual data points, which is represented by the spread between the best and worst mileage obtained by the test vehicle. Next, there is variation between the values of the individual retailers or treatments. This is referred to as between-treatment variation. In each experiment there is also a certain degree of background noise or error associated with the equipment, reading a dial, making a measurement, or recording a response. This is appropriately referred to as experimental error or, in this example, as within-treatment variation. The three types of variation described are shown graphically in Figure 7.2.

Table 7.2. Fuel Efficiency versus Retailer

Mileage (liters/100 km)					Mileage (mpg)				
Retailer_A	7.08	7.77	8.77	8.40	8.05	33.22	30.27	26.82	28.00	29.22
	9.33	7.99	8.32	8.66	7.92	25.21	29.44	28.27	27.16	29.70
Retailer_B	8.66	9.71	8.91	9.46	8.75	27.16	24.22	26.40	24.86	26.88
	9.77	8.88	10.33	8.92	9.98	24.07	26.49	22.77	26.37	23.57
Retailer_C	8.53	9.65	10.28	8.76	9.99	27.57	24.37	22.88	26.85	23.54
	8.13	8.79	9.40	8.42	10.89	28.93	26.76	25.02	27.93	21.60

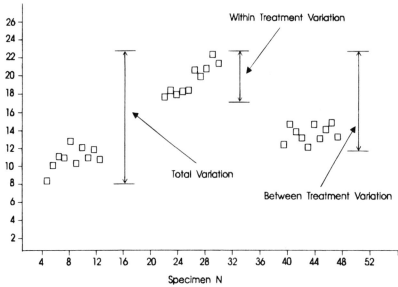

Figure 7.2. Types of variation found in data.

7.4 Variance

To do an ANOVA calculation a variance calculation must be carried out. The variance equation was introduced in Chapter 6 in Eq. (6.8). This equation is rewritten as Eq. (7.1).

$$s^2 = \frac{\sum_{i=1}^{n} (x_i)^2 - [(\sum x_i)^2/n]}{(n-1)} \tag{7.1}$$

7.5 Sum of Squares

Calculation of the variance in an ANOVA analysis is done in a slightly different manner than has been shown previously. To begin the variance calculation the numerator, or top half of the equation, is considered separately from the denominator, or bottom half of the equation. The numerator of Eq. (7.1) is referred to as the sum of squares, SS_x and is written as Eq. (7.2).

$$SS_x = \sum_{i=1}^{n} (x_i)^2 - \left[\frac{(\sum x_i)^2}{n} \right] \tag{7.2}$$

where SS_x = Sum of squares of treatment x

$\dfrac{(\Sigma x_i)^2}{n}$ = The overall variation observed in the data and is commonly referred to as the correction factor

$\displaystyle\sum_{i=1}^{n} (x_i)^2$ = Sum of squares of the individual data points.

7.6 Degrees of Freedom

Consider the denominator of Eq. (7.1). This portion of the variance formula is referred to as degrees of freedom of parameter x.

$$df_x = n_x - 1 \tag{7.3}$$

where n_x = the number of observations for treatment x.

In general terms the *degrees of freedom* can be described as the number of independent parameters associated with an experiment or factor (treatment). If a particular factor has a large number of degrees of freedom it implies that there is a significant amount of statistical information that can be obtained about the factor.

For example, if you were told about two physical characteristics of someone and then asked to identify the mystery person, the accuracy of the answer may be suspect as there would likely be numerous people who could fit this simplified description. If, on the other hand, you were provided with 10 physical attributes, the task of determining the identity of the mystery person becomes less prone to error. The same is true for degrees of freedom. The more degrees of freedom associated with the experiment or factor, the more confidence you, as an experimenter, will have about the results obtained.

7.7 Generation of a One-Way Analysis of Variance Table

The calculations required for a one-way ANOVA table can be performed by following a simple five-step process.

Step 1: Calculate the first half of the numerator or squares of the individual data points;

Step 2: Calculate the correction factor;

Step 3: Calculate the sums of squares for each treatment by combining the results from Steps 1 and 2;

Step 4: Calculate the degrees of freedom for each treatment in the experiment;

Step 5: Combine the results of steps 1 to 4, complete all remaining calculations, and summarize the results in the ANOVA Table. Organization of the data into this table makes the relationships between experimental factors and the results readily visible and is common for this analysis technique.

A blank one-way analysis of variance table is shown in Table 7.3.
Steps 1 through 5 are completed for the fuel efficiency experimentation.

Step 1: Determination of the Squares of Each Treatment

Squares for Retailer A

$$\sum_{i=1}^{n} (x_i)^2 = 7.08^2 + 7.77^2 + \cdots + 8.66^2 + 7.92^2$$

$$= 50.1 + 60.4 + \cdots + 75.0 + 62.7$$

$$= 680.61$$

Squares for Retailer B

$$\sum_{i=1}^{n} (x_i)^2 = 8.66^2 + 9.71^2 + \cdots + 8.92^2 + 9.98^2$$

$$= 74.99 + 94.28 + \cdots + 79.57 + 99.60$$

$$= 874.91$$

Squares for Retailer C

$$\sum_{i=1}^{n} (x_i)^2 = 8.53^2 + 9.65^2 + \cdots + 8.42^2 + 10.89^2$$

$$= 72.76 + 93.12 + \cdots + 70.9 + 118.59$$

$$= 869.31$$

Table 7.3. Blank ANOVA Table

Source	SS_x	df_x	MSS_x	$F_{Calculated}$	F_{Table}
Between					
Within					
Total					

Step 2: Calculation of the Correction Factor

$$\frac{(\Sigma \, x_i)^2}{n} = \frac{[\Sigma \, (7.08 + 7.77 + 8.77 + \cdots + 8.13 + 9.40 + 10.89)]^2}{30}$$

$$= \frac{(268.5)^2}{30}$$

$$= \frac{(72092.25)^2}{30}$$

$$= 2403.08$$

Step 3: Calculation of the Sum of Squares
Between Treatments

$$SS_{Between} = \frac{\Sigma \, (Retailer_A^2 + Retailer_B^2 + Retailer_C^2)}{10} - CF$$

$$= \frac{(82.29^2 + 93.37^2 + 92.84^2)}{10} - 2403.08$$

$$= \frac{24108.87}{10} - 2403.08$$

$$= 2410.89 - 2403.08$$

$$= 7.81$$

Total Sum of Squares

$$SS_{Total} = (Squares_{Treatment_A} + Squares_{Treatment_B} + Squares_{Treatment_C}) - CF$$

$$= (680.61 + 874.91 + 869.31) - 2403.08$$

$$= 2424.83 - 2403.08$$

$$= 21.75$$

Within Treatments

$$SS_{Within} = SS_{Total} - SS_{Between}$$

$$= 21.75 - 7.8$$

$$= 13.95$$

Step 4: Calculation of the Degrees of Freedom

$$df_{Within} = df_{Retailer_A} + df_{Retailer_B} + df_{Retailer_C}$$
$$= (10 - 1) + (10 - 1) + (10 - 1)$$
$$= 9 + 9 + 9$$
$$= 27$$

$$df_{Between} = \text{Number of treatments} - 1$$
$$= 3 - 1 = 2$$

$$df_{Total} = \text{Total number of data points} - 1$$
$$= 30 - 1 = 29$$

The degrees of freedom may also be calculated by subtracting the df calculated for the treatments from the total df:

$$df_{within} = df_{total} - df_{treatments}$$
$$= 29 - 2$$
$$= 27$$

Step 5: Generation of ANOVA Table

Three other parameters, mean sum of squares (MSS_x), $F_{Calculated}$, and F_{Table}, were also present in the ANOVA table shown in Table 7.3. To complete the ANOVA table these parameters must be calculated. The parameter MSS_x is equivalent to the variance (as described in Chapter 6) and is calculated in Table 7.4.

The calculated F statistic in a one-way ANOVA table is determined by dividing the calculated variance for a treatment by the variance calculated for the error term. In a one-way analysis of variance this parameter is represented by the within-

Table 7.4. Calculation of MSS_x for the Fuel Efficiency Experiment

Treatment	SS_x	df_x	MSS_x (SS_x/df_x)
Between	7.8	2	3.9
Within	13.95	27	0.52
Total	21.75	29	0.75

treatment calculations. Calculation of the F statistic for this example is shown in Eq. (7.4).

$$F_{\text{Calculated}} = \frac{\text{Variance}_{\text{Treatment}}}{\text{Variance}_{\text{Error}}}$$

$$\approx \frac{\text{MSS}_{\text{Treatment}}}{\text{MSS}_{\text{Error}}} \tag{7.4}$$

$$\approx \frac{\text{Mean sum squares}_{\text{Treatment}}}{\text{Mean sum squares}_{\text{Error}}}$$

Therefore, the F statistic for this example would be

$$F_{\text{Calculated}} = \frac{3.9}{0.52}$$

$$= 7.5$$

A theoretical F statistic for the chosen α-risk is obtained as described in Chapter 6. In this example there are two (2) degrees of freedom for the between-treatment variation and 27 for the within-treatment or error term. The degrees of freedom calculated for the treatment factor are listed along the top of the critical F statistic table and the degrees of freedom calculated for the error term are found down the vertical portion of the table as shown in Table 7.5. Since there are values listed for only 20 or 30, the value closest to the degrees of freedom for the treatment variable can be taken in its place.

Table 7.5. Determination of a Theoretical F Value

df$_{\text{error}}$ ⇓	1	2	3	4	5
1	161	20.0	216	225	230
2	18.5	19.0	19.2	19.2	19.3
3	10.1	9.55	9.28	9.12	9.01
4	7.71	6.94	6.59	6.39	6.26
5	6.61	5.79	5.41	5.19	5.05
6	5.99	5.14	4.76	4.53	4.39
7	5.59	4.74	4.35	4.12	3.97
8	5.32	4.46	4.07	3.84	3.69
9	5.12	4.26	3.86	3.63	3.48
10	4.96	4.10	3.71	3.48	3.33
12	4.75	3.69	3.49	3.26	3.11
15	4.54	3.68	3.29	3.06	2.90
20	4.35	3.49	3.10	2.87	2.71
30	4.17	**3.32**	2.92	2.69	2.53

An F statistic has a standard form in which it is written. This format indicates the α-risk assumed, 95% in this example, and the degrees of freedom for both the treatment factor (2) and error term (27). This format is as follows:

$$F_{2,27(0.05)} = 3.32$$

The finished one-way ANOVA table, containing all of the calculations, is shown in Table 7.6.

The above analysis has shown that the experimental F statistic for the treatment factor is greater than the theoretical F value obtained from the literature tables. This would suggest that the differences in fuel efficiency could very well be assigned to a particular retailer. This would mean that the null hypothesis, which states there is no difference between retailers, should be rejected in favor of our alternate hypothesis, which says that the observed changes in fuel economy can be assigned to a gasoline retailer.

In practical terms, however, labeling factors as significant or nonsignificant based on the F statistic is not as straightforward as it appears. In reality an F statistic is only an indication that the observed differences are larger than what could rationally be attributable to random noise. Changing the risk level will change the theoretical F statistic. For example, if confidence level of 0.10 was used, the literature value for the F ratio for this example would decrease to 2.51. If, on the other hand, a level of 0.01 had been used the critical F statistic would increase to 5.49.

It is possible to make a factor statistically significant or insignificant by using a different α-risk level. Suffice it to say that for most practical experimental situations, the main function of the F statistic is to establish a threshold value with which to guide the experimenter in making a decision about a factor and to avoid making a Type I or II error as previously discussed in Chapter 6.

7.8 Two-Way Analysis of Variance

The analysis of variance technique can be used to study any number of factors. To investigate a more complicated situation than was done in the previous example, an additional factor will be added to the matrix. For this second example, seasonal effects will be investigated as well as gasoline retailers, and the previous exercise will be repeated. This procedure is referred to as a two-way analysis of variance.

Table 7.6.   ANOVA Table with All Parameters Calculated

Source	SS_x	df_x	MSS_x	$F_{Experimental}$	F_{Table}
Between	7.80	2	3.90	7.5	3.32
Within	13.95	27	0.52		
Total	21.75	29	0.75		

The null and alternative hypothesis for this second experiment would be stated as follows:

- Null hypothesis: Measured differences in fuel efficiency cannot be attributed to either gasoline retailer or season

  $H_0: \mu_1 = \mu_2$

- Alternate hypothesis: Measured differences in fuel efficiency can be attributed to either gasoline retailer or season

  $H_a: \mu_1 \neq \mu_2$

Table 7.7 breaks down fuel efficiency into retailer and the season of the purchase. Three categories, Spring/Fall (March to April, September to October), Winter (November to February), and Summer (May to August) have been used to define the season. Values are presented in liters/100 km.

A two-way analysis of variance is set up similar to the one-way analysis. The five steps performed previously would be repeated using the new data.

Step 1: Determination of Squares

Squares Retailer$_A$

$$\sum_{i=1}^{n} (x_i)^2 = 8.77^2 + 8.14^2 + 9.08^2 + \cdots + 10.33^2 + 9.15^2 + 9.45^2$$

$$= 1718.72$$

Squares Retailer$_B$

$$\sum_{i=1}^{n} (x_i)^2 = 9.11^2 + 8.25^2 + 8.67^2 + \cdots + 9.98^2 + 8.88^2 + 9.92^2$$

$$= 1702.17$$

Squares Retailer$_C$

$$\sum_{i=1}^{n} (x_i)^2 = 8.23^2 + 7.55^2 + 7.62^2 + \cdots + 10.89^2 + 9.99^2 + 9.72^2$$

$$= 1730.38$$

Table 7.7. Fuel Efficiency versus Retailer and Season Purchased

	Spring/Fall			Summer			Winter		
Retailer$_A$	8.77	8.33	9.55	7.65	7.98	7.98	10.07	10.86	9.45
	8.14	8.99		8.03	7.77		10.56	10.33	
	9.08	9.02		8.31	8.57		10.32	9.15	
Retailer$_B$	9.11	8.66	8.82	8.72	8.21	8.55	9.91	10.33	9.92
	8.25	8.91		8.36	8.14		9.46	9.98	
	8.67	8.75		8.92	8.26		9.77	8.88	
Retailer$_C$	8.23	8.53	8.60	9.03	8.03	8.01	10.65	9.40	9.79
	7.55	8.28		9.43	9.40		10.76	10.89	
	7.62	8.42		8.70	7.99		10.13	9.99	

Two factors are now being evaluated, so the sums associated with both parameters need to be calculated. For the gasoline retailer this is accomplished by obtaining the horizontal sum of the individual data points for each retailer. For seasonal effects the vertical sum of the individual data points, according to season, is found.

Sums of Retailer Effects

$$\text{Retailer}_A = 8.77 + 8.14 + 9.08 + \cdots + 10.33 + 9.15 + 9.45$$

$$= 188.91$$

$$\text{Retailer}_B = 9.11 + 8.25 + 8.67 + \cdots + 9.98 + 8.88 + 9.92$$

$$= 188.58$$

$$\text{Retailer}_C = 8.23 + 7.55 + 7.62 + \cdots + 10.89 + 9.99 + 9.79$$

$$= 189.43$$

Sum of Seasonal Effects

$$\text{Spring/Fall} = 8.77 + 8.14 + 9.08 + \cdots + 8.28 + 8.42 + 8.60$$

$$= 180.28$$

$$\text{Summer} = 7.65 + 8.03 + 8.31 + \cdots + 9.40 + 7.99 + 8.01$$
$$= 176.04$$

$$\text{Winter} = 10.07 + 10.56 + 10.32 + \cdots + 10.89 + 9.99 + 9.79$$
$$= 210.60$$

Step 2: Calculation of the Correction Factor

$$\frac{(\Sigma\ x_i)^2}{n} = \frac{[\Sigma\ (8.77 + 8.14 + 9.08 + \cdots + 10.89 + 9.99 + 9.79)]^2}{63}$$

$$= \frac{(566.92)^2}{63} = \frac{321398.29}{63}$$

$$= 5101.56$$

Step 3: Calculation of the Sum of Squares
Sum of Squares for Gasoline Retailer

$$SS_{Retailer} = \frac{Retailer_A^2 + Retailer_B^2 + Retailer_C^2}{21} - CF$$

$$= \frac{188.91^2 + 188.58^2 + 189.43^2}{21} - 5101.56$$

$$= 5101.58 - 5101.56$$

$$= 0.02$$

Sum of Squares for Season

$$SS_{Season} = \frac{(Spring/Fall)^2 + Winter^2 + Summer^2}{21} - CF$$

$$= \frac{180.28^2 + 176.04^2 + 210.6^2}{21} - 5101.56$$

$$= 5135.4 - 5101.56$$

$$= 33.84$$

Sum of Squares: Total

$$SS_T = (Squares\ Retailer_A + Squares\ Retailer_B + Squares\ Retailer_C) - CF$$

$$= (1718.7 + 1702.2 + 1730.4) - 5101.56$$

$$= 5151.3 - 5101.56$$

$$= 49.74$$

Sum of Squares: Error

$$SS_{Error} = SS_{Total} - SS_{Season} - SS_{Retailer}$$
$$= 49.74 - 33.84 - 0.02$$
$$= 15.88$$

Step 4: Calculation of the Degrees of Freedom

$$df_{Retailer} = \text{Number of treatments} - 1$$
$$= 3 - 1 = 2$$
$$df_{Season} = \text{Number of treatments} - 1$$
$$= 3 - 1 = 2$$
$$df_{Total} = \text{Number of data points} - 1$$
$$= 63 - 1 = 62$$
$$df_{Error} = df_{Total} - df_{Retailer} - df_{Season}$$
$$= 62 - 2 - 2$$
$$= 58$$

Step 5: Generation of ANOVA Table

Three remaining parameters, MSS_x, experimental F statistic, and the theoretical F statistic, in the ANOVA table are calculated below. First the calculation of MSS_x is shown in Table 7.8.

Table 7.8. Calculation of MSS_x for the Second Fuel Efficiency Experiment

Treatment	SS_x	df_x	MSS_x (SS_x/df_x)
Retailer	0.02	2	0.01
Season	33.84	2	16.92
Error	15.88	58	0.27
Total	49.74	62	0.80

Table 7.9. Calculation of the F Statistic for the Seasonal Fuel Efficiency Experiment

Treatment	MSS_x (SS_x/df_x)	F calculated ($MSS_{Treatment}/MSS_{Error}$)
Retailer	0.01	0.04
Season	16.92	62.67
Error	0.27	
Total	0.79	

7.9 Determination of F Statistic

As shown previously the experimental F statistic for a treatment is determined by dividing the calculated variance for the treatment by the error term variance. Calculation of the experimental F statistic is shown in Table 7.9.

A critical F ratio for the chosen α-risk is obtained as previously and is shown in Table 7.10. In this example there are 2 degrees of freedom associated with each treatment and 58 for the error term.

The finished ANOVA table containing all of the calculations is summarized in Table 7.11.

Table 7.10. Determination of Theoretical F Statistic

df_{error} ⇩	1	2	3	4	5
10	4.96	4.10	3.71	3.48	3.33
12	4.75	3.89	3.49	3.26	3.11
15	4.54	3.68	3.29	3.06	2.90
20	4.35	3.49	3.10	2.87	2.71
30	4.17	3.32	2.92	2.69	2.53
60	4.00	**3.15**	2.76	2.53	2.37

Table 7.11. Completed ANOVA Table for Seasonal Fuel Efficiency Experiment

Treatment	SS_x	df_x	MSS_x	$F_{Calculated}$	F_{Table}
Retailer	0.02	2	0.01	0.04	3.15
Season	33.84	2	16.92	62.67	
Error	15.88	58	0.27		
Total	49.74	62	0.80		

The large experimental F statistic associated with seasonal effects and the insignificant theoretical F statistic calculated for the retailer suggest that seasonal effects was a main contributor to variations in fuel economy. This also means that the null hypothesis, which states there is no difference between retailers or seasonal effects, should be rejected in favor of our alternate hypothesis, which says that the observed changes in fuel economy could be assigned to the gasoline retailer or seasonal effects.

In the next chapter the ANOVA technique is applied to multitreatment, three or more, systems using a specific type of statistical experiment called screening designs.

8

Screening Experiments

"Statistics are like alienists—they will testify for either side."
Fiorello La Guardia, 1882–1947, American politician, mayor of New York

8.1 What Are Screening Experiments?

Analysis of variance has been used to analyze the data from an experiment in which two or three factors were examined. Rarely is a process so thoroughly understood that only one or two factors will need to be investigated. More commonly, there will be closer to 10 or 20 factors that could be investigated. Looking at a process using a one- or two-way analysis of variance would not be an efficient use of time in most industrial situations.

Is the experimenter committed to looking at all the possible combinations of the factors being studied or is there another method that decreases the number of experiments needed? Fortunately, statisticians have developed a series of methodologies that can be used to investigate a large number of factors. These methods or designs are called by many different names. Orthogonal arrays and Plackett–Burman designs are two types of designs that can be used for screening experiments.

8.2 Orthogonal Arrays

Orthogonal arrays are used extensively by followers of Dr. Taguchi's design methodology. Some of the most common orthogonal arrays in use are an L_4, L_8, L_{16}, and L_{32}. Designations for orthogonal arrays include the letter "L" first then the subscript

Table 8.1. L_4 Orthogonal Array

		Factors	
Contrast	A	B	C
1	1	1	1
2	1	2	2
3	2	1	2
4	2	2	1

number second. The subscript after the L denotes the number of trials that must be executed in a given design. For example, in an L_8 eight trials would be required to complete the experiment. In classical statistics a trial is also referred to as a "contrast."

Orthogonal arrays are also generally referred to as balanced designs. A balanced design means that each factor in the array will be compared to all other factors in a design an equal number of times. Consider the array shown in Table 8.1, an L_4.

Table 8.2 summarizes how many times each factor is compared to each other factor in the L_4 array. This table shows that each factor in the array (i.e., A, B and C) will be compared to each other factor in the array once at each level. The same process could be done for an L_8 and it would be determined that each factor is compared to each other factor twice at each setting.

Of the numerous screening designs available for use by an experimenter L_8, Table 8.3, and L_{16}, Appendix B, arrays are two of the most frequently used. In these designs the experimenter can evaluate between 3 to 7 and 4 to 15 factors, respectively.

8.3 Confounding

There has been no mention about the underlying structure of an experimental array. The underlying structure of a design is generally referred to as a design's confounding

Table 8.2. Comparison of Factors in an L_4 Array

	Vs A		Vs B		Vs C	
	Level 1	Level 2	Level 1	Level 2	Level 1	Level 2
Factor A						
Level 1	—	—	Trial 1	Trial 2	Trial 1	Trial 2
Level 2	—	—	Trial 3	Trial 4	Trial 4	Trial 3
Factor B						
Level 1	Trial 1	Trial 3	—	—	Trial 1	Trial 3
Level 2	Trial 2	Trial 4	—	—	Trial 4	Trial 2
Factor C						
Level 1	Trial 1	Trial 4	Trial 1	Trial 4	—	—
Level 2	Trial 2	Trial 3	Trial 3	Trial 2	—	—

Table 8.3. L_8 Array

Contrast	Factors						
	A	B	C	D	E	F	G
1	1	1	1	1	1	1	1
2	1	1	1	2	2	2	2
3	1	2	2	1	1	2	2
4	1	2	2	2	2	1	1
5	2	1	2	1	2	1	2
6	2	1	2	2	1	2	1
7	2	2	1	1	2	2	1
8	2	2	1	2	1	1	2

pattern or alias structure. This means that there are factor combinations associated with each column in an array other than the main factor, which has been directly assigned to a particular column by the experimenter. Each column in an experimental design has a series of multifactor interactions associated with it that have been clearly defined when the design was formulated.

8.3.1 Interactions

What is meant by an interaction? An interaction describes a specific type of relationship between two or more process factors. An interaction is said to exist when changes in one factor cause changes in the response of another factor. In other words, the response generated by changes of one factor will be directly affected by the changes made to the factor(s) with which it interacts. Noninteracting and interacting data are shown graphically in Figures 8.1 to 8.3.

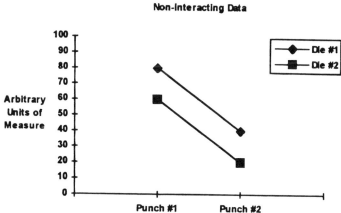

Figure 8.1. Noninteracting data.

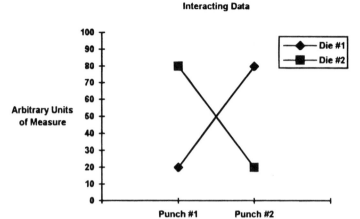

Figure 8.2.   Interacting data, Type 1.

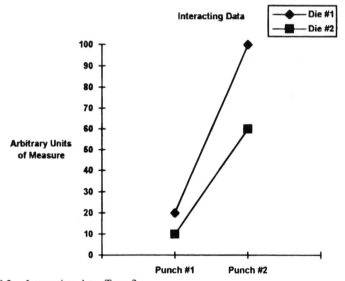

Figure 8.3.   Interacting data, Type 2.

Different levels of complexity of interactions exist in nature, but by far the most common interaction will be between two independent factors as shown above. This type of interaction is called a 2-factor interaction. Interactions between three or more factors are much less common and require significantly greater effort on the part of the experimenter to interpret. In general, interactions are not investigated until there is evidence, such as nonreproducible results or large error terms, to suggest that they actually exist.

When a main factor occupies a column that also has a multifactor interaction associated with it, the effect of the multifactor interaction cannot be distinguished from the main effect without additional experimentation or mathematical manipulation.[1] Knowledge of where interactions may occur in the design is critical to being able to ultimately understand the system under investigation once results have been generated.

8.3.2 Confounding Patterns

Confounding patterns can be developed manually and a simple example of how this can be done follows. To make the determination of the confounding patterns or alias structures easier, the "1s" and "2s" normally used in the array will be replaced with "−1" and "+1," respectively. A cross product between two factors, A and B (denoted by a × b) for example, is obtained by multiplying the signs of each factor together and will be representative of the interaction between factors A and B. The cross product of positive and negative signs will result in a negative sign and the cross product of two positive or two negative signs culminates in a positive sign.

This process is illustrated for the simple case of two factors in Table 8.4. Increasing the main factors to three, the confounding pattern for this array would be as shown in Table 8.5.

It was quite easy to obtain the confounding pattern in these two examples, but as the numbers of factors continue to increase generating and deciphering the confounding pattern will not be such an easy task. Widespread introduction of the personal computer coupled with the appropriate statistical software has eliminated the need to know how to generate confounding patterns on a routine basis. Confounding patterns can also be obtained by consulting a table containing the alias structure for use with the design being considered. Common confounding patterns or alias structures for selected arrays and main factor combinations are described in Appendix C.

Table 8.4.   Cross Product of Factors A and B

| | Factors | | |
Contrast	A	B	axb
1	−1	−1	+1
2	−1	+1	−1
3	+1	−1	−1
4	+1	+1	+1

[1] A technique for identifying an interaction effect is outlined in Wheeler, D.J. "Tables of Screening Designs." Statistical Process Controls, Inc., Knoxville, Tennessee, 1988, pg. 27–31

Table 8.5. Simple Confounding Patterns for Three Main Factors

				Factors			
Contrast	A	B	C	(axb)	(axc)	(bxc)	(axbxc)
1	−1	−1	−1	+1	+1	+1	−1
2	−1	−1	+1	+1	−1	−1	+1
3	−1	+1	−1	−1	+1	−1	+1
4	−1	+1	+1	−1	−1	+1	−1

8.4 Resolution

Experimental designs can also be described with respect to their resolution. The *resolution* of an experiment permits the experimenter to define how many independent factors will be free from interference from interaction effects. The greater the resolution of a design the lower the degree of confounding. However, there is also a corresponding increase in the number of trials required to study a given number of factors.

Regardless of the number of trials or factors studied, three types of resolution are of interest in this text:

Resolution III   A saturated (i.e., all columns in the design are filled with a process factor) fractional factorial design that does not confound main effects with other main effects but does confound 2-factor interactions with main effects;

Resolution IV   An unsaturated fractional factorial design (i.e., not all columns in the design are filled with process factors) that does not confound main effects with 2-factor interactions but does confound 2-factor interactions with other 2-factor interactions. Three-way factor interactions are confounded with main effects;

Resolution V   A full factorial design (i.e., only the minimal number of columns in the design are filled with process factors) that does not confound main factor effects with other main factor effects or with 2-factor interactions. Two-factor interactions are also not confounded with one another. All three-way or greater interactions are assumed to be zero.

8.5 Selecting a Design

A general rule for choosing one experimental design over another in the screening stages of experimentation is to always use the smallest array that is capable of handling the number of factors selected for study. For example, if three factors were of interest, an L_4 would most likely be used. If there were six factors, an L_8 would be needed. Other considerations such as the goals and objectives of the investigation, availability of production time, budget for the trials, known existence of interactions, and the severity of the problem will also influence the choice of an array. A common guideline is to use 20–30% of the total number of trials available/allowed as screening experiments.

As part of selecting an experimental design the experimenter needs to determine what resolution is acceptable for the experiment. Generally, during the screening stage either Resolution III or IV designs are acceptable. Table 8.6 summarizes the relationship between the number of factors being studied, the size of the experiment needed, the resolution that will result from each combination of factors and experimental design size, and the fraction of the total number of experimental trials that would be required if all factor combinations were investigated.

8.6 Selecting Response Variables

After running each trial in a designed experiment a process value of some kind needs to be recorded, an analytical test needs to be performed, or an evaluation of some other type needs to be made. Variables such as these are called response variables or quality characteristics. An ideal response variable will be an absolute measure of the performance of the process or product and should be continuous whenever possible.

Each designed experiment should measure at least two or three different response variables or quality characteristics. It is very likely that if only one response variable is studied a different response variable may be adversely affected. To decide which responses need to be measured, the question "Does it matter to the customer?" should be asked. If it is not important to the customer then it likely does not need to be measured for the experiment. However, if the customer is inclined to notice a change in the product or the process parameter (e.g., tensile strength or impact properties) it would be prudent to evaluate this parameter.

Table 8.6. Relationship between Factors, Resolution, and Number of Trials Required[a]

Trials Required in an Array	Factors in an Array													
	2	3	4	5	6	7	8	9	10	11	12	13	14	15
	V[b]	III												
4	Full	1/2												
		V	IV	III	III	III								
8		Full	1/2	1/4	1/8	1/16								
			V	V	IV	IV	IV	III	III	III	III	III	III	III
16			Full	1/2	1/4	1/8	1/16	1/32	1/64	1/128	1/256	1/512	1/1024	1/2048
				V	IV	IV	IV	IV	IV	IV	IV	IV	IV	IV
32				Full	1/2	1/4	1/8	1/16	1/32	1/64	1/128	1/256	1/512	1/1024

[a]Data reprinted with permission from Stat-Ease from the Design Ease® V2.0 experimental design software.
[b]III, IV, V, resolution numbers.

8.7 Assigning Factors to a Design

It is not normally necessary to assign a factor to a specific column in an array because the order of running the trials in an experiment is typically randomized. Randomization is done to minimize the influence of experimental error and processes that drift with time. For example, if the trials in an design were done in sequential order the results would be biased (either positively or negatively depending on the response) toward the experiments done last due to the drift in the process. Erroneous results would be obtained and the process could not be optimized. Conversely, if the order of experimentation is randomized, the error that is introduced into the results will be distributed relatively equally over all trials and not just the last one or two trials.

To randomize the trials in a design the table of random numbers found in Table 3 of Appendix D can be used.[2] To use this table, simply chose any one of the nine columns in the table and follow it down until a number in that column corresponds to one of the trial numbers in the array being randomized. This would then be the first experiment conducted. This procedure is repeated until each trial had been found and a complete run order is obtained. If the end of the column is reached the search for the next number is simply continued in the adjacent column.

For example, if column one were used to randomize the trials in an L_4 array, the run order would be 2, 4, 1, and 3. If, for example, the trials of an L_8 array needed to be randomize, and the search for a random order was begun in column 5, the run order of the trial would be 8, 7, 2, 5, 4, 6, 3, and 1.

However, sometimes there are factors that are quite difficult to change or the effects of a change in this factor take a long time to show (e.g., temperature or materials); the following column assignment technique can then be used. Consider the first column in an L_8 array. Only one change in the factor level is needed. Column 2 requires three changes, column 3 requires three changes, column 4 requires seven changes, column 5 requires six changes, column 6 requires four changes, and column 7 requires five changes. Difficult to change variables are typically put in the first two columns of a two-level array and are referred to as Group 1 factors. Factors that are less difficult to change or in which the effects of change are observed immediately (e.g., pressure or time) are called Group 2 factors and are placed in any one of the remaining columns. As discussed previously, each column found in a design consists of both main effects and multifactor interactions. If an interaction is suspected between two factors and both factors responsible for the interaction must be looked at, then no main factor should be placed in the column where the interaction is located. This column should be left open.

---

[2] Computer software packages, such as Design-Ease, automatically randomize experiments for the experimenter.

8.8 Choosing Factor Ranges

Whenever practical the difference between levels 1 and 2 should approximate the known operating limits or molding window of the process. It is important to use as wide a range as practically possible, as this magnifies the effect of the response or quality characteristic being investigated and minimizes the potential effect of experimental error. If the range is made too small it is possible that the factor could appear insignificant when it is not. Conversely, if the range is made too wide it will dominate the experiment and appear to be the only significant factor in the experiment when it really is not. Each situation must be dealt with on an individual basis and personal expertise and judgment should be used to set safe and practical limits.

8.9 Determining the Reasonableness of the Ranges Selected

Constructing a Table of Reasonableness is one approach in developing reasonable ranges for a factor. This table is simply a listing of each trial combination to be run in the experiment. This table provides a visual review of the various process conditions prior to running them in production. If some combinations are considered to be unacceptable or are known to be troublesome, changes can be made prior to actually conducting a trial. This helps prevent wasting valuable production time. When process knowledge is limited, the design team's "best guess" must be used.

Consider the factor data and their ranges found in Table 8.7. Using an L_8 for the design in this experiment the resulting Table of Reasonableness is shown in Table 8.8.

Each trial found in the Table of Reasonableness should be reviewed by the process operators and production people before proceeding to the manufacturing trials. If any combination of factors is considered not acceptable, because of past experience, for example, settings for the factors should be adjusted accordingly. Adjustments are made to the unreasonable factor level(s) until there is consensus among the design team that the trials being considered can be run safely.

Table 8.7. Factors and Ranges for a
New Epoxy Compound

Factor	Range
(A) Cure temperature (°C)	225 vs 275
(B) Filler (%)	5–10
(C) Resin (Type)	A vs B
(D) Diluent (%)	2 vs 6
(E) Mixing speed (rpm)	1000 vs 2000

Table 8.8. Table of Reasonableness for the Epoxy Experiment

Trial Number	Factor				
	A (°C)	B (%)	C (Matl)	D (%)	E (rpm)
1	225	5	A	2	1000
2	225	5	A	6	2000
3	225	10	B	2	1000
4	225	10	B	6	2000
5	275	5	B	2	2000
6	275	5	B	6	1000
7	275	10	A	2	2000
8	275	10	A	6	1000

8.10 Sample Size

Another question that needs to be considered is the number of samples (or measurements) that should be taken for each trial. Common sense suggests that it is generally advisable to take more than one measurement for each trial, however, taking a large number of measurements does not always improve the answer obtained substantially. The reason is that the precision of a measurement is proportional to $1/\sqrt{n}$, where n is the number of samples tested.

An illustration of the increase in precision is shown in Table 8.9.

If a large change in the process mean is of interest, regardless of the size of the standard deviation, then a small sample (up to 5) set will likely be sufficient for most experiments. However, if it is necessary to detect very small process changes and achieve this with any degree of accuracy, much larger sample sizes are required (greater than 25).

8.11 Determination of Sample Sizes

Appropriate sample sizes can also be calculated when the standard deviation, the size of the change in process mean desired, and the magnitude of the α- and β-

Table 8.9. Improvements in Precision as a Function of the Number of Samples Tested

n	$1/\sqrt{n}$	Improvements in precision vs only one measurement (%)
1	1.00	—
5	0.45	65
25	0.20	80
100	0.10	90

risks assumed for the experiment are known. For industrial experiments α- and β-risks typically used are 0.05 and 0.10, respectively. Selected values of $t_{\alpha/2}$ and t_β are listed in Appendix D, Table 2.

The formula[3] used in this calculation is

$$n = \left\{ \frac{[(t_{\alpha/2} + t_\beta) \times \sigma]}{\Delta d} \right\}^2 \tag{8.1}$$

where n = number of samples required

$t_{\alpha/2}$ = area of the t distribution representing 1/2 of the α-risk

t_β = area of the t distribution corresponding to the β-risk

σ = standard deviation of the process measurement

Δd = desired difference to be detected.

How many samples would need to be tested if a difference of 10 psi needed to be detected in a process with a standard deviation of 5 psi? The sample size used to calculate the standard deviation was greater than 50. Acceptable α- and β-risks assumed for this example are 0.05 and 0.10, respectively.

The first step in this calculation will be to determine the values of the different parameters in the formula. The appropriate values for $t_{\alpha/2}$ and t_β are determined in much the same way as obtaining a theoretical F statistic, except Table 4 in Appendix D is used for this purpose. The column for the selected $t_{\alpha/2}$ value is located (i.e., 0.025 since an α-risk of 0.05 was chosen) in this table and then followed down until the row corresponding to one less than the number of samples measured is found. This would then be the number used for this parameter. This is illustrated in Table 8.10.

Table 8.10. Determining $t_{\alpha/2}$

Sample Size	$t_{0.025}$
1	12.706
5	2.571
10	2.228
15	2.131
20	2.086
25	2.060
∞	1.960

[3] Rickmers, A. D. and Todd, H. N., "Statistics: An Introduction," pg. 125–137, McGraw-Hill, NY, 1967.

The same table and procedure is used to find a value for t_β. *In this example t_β* would be 1.282. Therefore using $t_{\alpha/2} = 1.960$

$$t_\beta = 1.282$$

$$\sigma = 5$$

$$\Delta d = 10$$

In Eq. (8.1) the required sample size would be as follows:

$$n = \left\{ \frac{[(1.960 + 1.282) \times 5]}{10} \right\}^2$$

$$= \left[\frac{(3.242 \times 5)}{10} \right]^2$$

$$= (1.621)^2$$

$$= 2.628$$

This means that three measurements would be sufficient to detect the required difference under the stated conditions.

IN EACH
TRIAL

CHAPTER

9

ANOVA for Screening Experiments

"I always find that statistics are hard to swallow and impossible to digest. The only one I can ever remember is that if all the people who go to sleep in church were laid end to end they would be a lot more comfortable."

Mrs. Robert A. Taft, wife of an American politician

9.1 Case Study: Batting Average

A concerned batting coach was looking at ways he could improve the batting average of one of his players. He decided to experiment with some of the factors he felt might affect this player's batting technique. The following factors and settings were of interest to the coach:

- Baseball Bat
    Heavy weight vs light weight (Factor A)
    Short length vs long length (Factor B)
    Wood vs aluminum (Factor C)
- Distance to Pitcher
    Front of plate vs back of plate, Position P_1 (Factor D)
- Distance to the Plate
    One foot vs two feet, Position P_2 (Factor E)
- Footwear
    No spikes vs spikes (Factor F)

Table 9.1. Table of Reasonableness for baseball experiment

<table>
<tr><td colspan="8" align="center">Table of Reasonableness</td></tr>
<tr>
<td></td>
<td>A
Bat
Wt.</td>
<td>B
Bat
Length</td>
<td>C
Bat
Matl.</td>
<td>D
Distance
to Pitcher</td>
<td>E
Distance
to Plate</td>
<td>F
Footwear</td>
<td>G
Stance</td>
</tr>
<tr><td>1</td><td>Heavy</td><td>Short</td><td>Wood</td><td>Front</td><td>One</td><td>No</td><td>Right</td></tr>
<tr><td>2</td><td>Heavy</td><td>Short</td><td>Wood</td><td>Back</td><td>Two</td><td>Yes</td><td>Left</td></tr>
<tr><td>3</td><td>Heavy</td><td>Long</td><td>Alnm</td><td>Front</td><td>One</td><td>Yes</td><td>Left</td></tr>
<tr><td>4</td><td>Heavy</td><td>Long</td><td>Alnm</td><td>Back</td><td>Two</td><td>No</td><td>Right</td></tr>
<tr><td>5</td><td>Light</td><td>Short</td><td>Alnm</td><td>Front</td><td>Two</td><td>No</td><td>Left</td></tr>
<tr><td>6</td><td>Light</td><td>Short</td><td>Alnm</td><td>Back</td><td>One</td><td>Yes</td><td>Right</td></tr>
<tr><td>7</td><td>Light</td><td>Long</td><td>Wood</td><td>Front</td><td>Two</td><td>Yes</td><td>Right</td></tr>
<tr><td>8</td><td>Light</td><td>Long</td><td>Wood</td><td>Back</td><td>Two</td><td>No</td><td>Left</td></tr>
</table>

- Stance
  Right vs left (Factor G)

These factors were incorporated into an L_8 design. The Table of Reasonableness generated by assigning the chosen factors to this array is shown in Table 9.1.

The slumping ball player used each combination suggested to him by the coach for a consecutive four game stretch. During the four games the player's batting average was calculated to measure how effective each combination had been. Table 9.2 summarizes the batting averages achieved over the 8 weeks the experiment was run.

An analysis of variance calculation for a multifactor experiment is done in a slightly different manner than shown previously. The steps required to complete the analysis of variance calculation are described in Figure 9.1.

As indicated in Figure 9.1 and as with two-way analysis of variance calculation the totals of each trial need to be determined first. This value is obtained by computing the horizontal sum of the individual data points in each trial. Once totals have been

Table 9.2. Batting Averages during Trials

Week Number	Game 1	Game 2	Game 3	Game 4
1	.210	.225	.210	.220
2	.260	.268	.255	.268
3	.350	.360	.348	.365
4	.300	.298	.305	.303
5	.175	.188	.180	.185
6	.360	.365	.370	.368
7	.500	.485	.490	.503
8	.420	.413	.428	.410

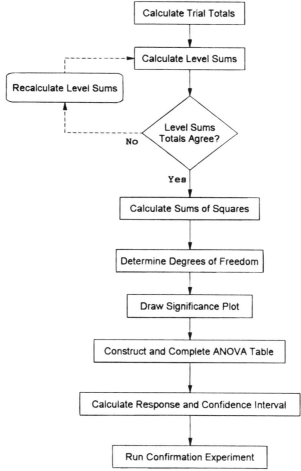

Figure 9.1. Flow diagram for carrying out an ANOVA calculation for a screening experiment.

obtained for each trial they are then totaled to obtain an overall experimental total. These calculations are shown in Table 9.3.

9.2 Calculation of the Level Sums

Table 9.4 summarizes the trials that need to be added together to obtain the Level 1 and Level 2 totals for each factor. This parameter is needed to calculate the sums of squares.

Consider factor C. In an L_8 array factor C is set at level 1 in trials 1, 2, 7, and 8. Therefore, calculation of the C_1 level sum is accomplished by adding together the totals for a each of these trials as shown in Table 9.5.

Table 9.3. Trial Sums and Experimental Total

Week Number	Game 1		Game 2		Game 3		Game 4	Total
1	.210	+	.225	+	.210	+	.220	0.865
2	.260	+	.268	+	.255	+	.268	1.051
3	.350	+	.360	+	.348	+	.365	1.423
4	.300	+	.298	+	.305	+	.303	1.206
5	.175	+	.188	+	.180	+	.185	0.728
6	.360	+	.365	+	.370	+	.368	1.463
7	.500	+	.485	+	.490	+	.503	1.978
8	.420	+	.413	+	.428	+	.410	1.671
							Experiment total =	10.385

Table 9.4. Trial Combinations for Calculating the Level Sums for Each Factor in an L_8 Array

Factor	Level 1 Trial Numbers	Level 2 Trial Numbers
A	1, 2, 3, 4	5, 6, 7, 8
B	1, 2, 5, 6	3, 4, 7, 8
C	1, 2, 7, 8	3, 4, 5, 6
D	1, 3, 5, 7	2, 4, 6, 8
E	1, 3, 6, 8	2, 4, 7, 9
F	1, 4, 5, 8	2, 3, 6, 7
G	1, 4, 6, 7	2, 3, 5, 8

Table 9.5. Level Sums Calculation for Factor C at Level 1

Trial Number	A	B	C	D	E	F	G	Total
1	1	1	**1**	1	1	1	1	**0.865**
2	1	1	**1**	2	2	2	2	**1.051**
3	1	2	2	1	1	2	2	—
4	1	2	2	2	2	1	1	—
5	2	1	2	1	2	1	2	—
6	2	1	2	2	1	2	1	—
7	2	2	**1**	1	2	2	1	**1.978**
8	2	2	**1**	2	2	1	2	**1.671**
					Level sum total for C_1			**5.565**

Table 9.6. Level Sums for Factor F at Level 2

Trial Number	A	B	C	D	E	F	G	Total
1	1	1	1	1	1	1	1	—
2	1	1	1	2	2	**2**	2	**1.051**
3	1	2	2	1	1	**2**	2	**1.423**
4	1	2	2	2	2	1	1	—
5	2	1	2	1	2	1	2	—
6	2	1	2	2	1	**2**	1	**1.463**
7	2	2	1	1	2	**2**	1	**1.978**
8	2	2	1	2	2	1	2	
					Level sum total for F_2			**5.915**

Determining the level sum for an additional factor, F, at level 2, trials number 2, 3, 6, and 7 would need to be totaled. This calculation is illustrated in Table 9.6.

9.3 Level Sums Table

Once the level sums for each factor have been calculated, they should be organized into a Level Sums Table as illustrated in Table 9.7. The totals for each factor are also calculated and recorded in this table.

To determine if there has been a calculation error up to this point, the sum of level 1 and level 2 is obtained. If this sum equals the total for the experiment, the mathematics have been done correctly. A visual check of the data in Table 9.7 indicates that this is true, so it is safe to assume that no errors have been made.

A second calculation check can be performed to confirm the level sum calculations. This secondary check requires the calculation of the vertical sum of all the responses for level 1 and level 2. Once calculated, these two numbers are then

Table 9.7. Level Sums Table

Factor	Level 1 Total	Level 2 Total	Total Level 1 + Level 2
Bat weight	4.545	5.840	10.385
Bat length	4.107	6.278	10.385
Bat material	5.565	4.820	10.385
Distance to pitcher	4.994	5.391	10.385
Distance from plate	5.422	4.963	10.385
Footwear	4.470	5.915	10.385
Stance	5.512	4.873	10.385

summed together to obtain an overall level sums total. Next the sum of each trial, 10.385 in this example, is multiplied by the number of trials (i.e., 8) carried out in the experiment. If all level sums calculations were done correctly, these two numbers should be the same. If they are not, a calculation error has been made and all additions should be checked. The results of this secondary calculation check are summarized in Table 9.8 below.

If no checks are made and there was a calculation error, it would be carried through the rest of the ANOVA analysis leading to erroneous results. When performing these calculations by hand, perform all of the suggested calculation checks.

9.4 Sums of Squares

The next operation to be performed is the sums of squares (SS_x) calculation. This parameter was first outlined in Chapter 7, Eq. (7.4). However, the sums of squares calculation used in a multifactor designed experiment is not exactly the same as this calculation. It is necessary to incorporate the results from each level of a factor into the calculation. The modified sums of squares calculation in a two-level designed experiment is as follows:

$$SS_x = \left(\frac{\text{Level Sums}^2_{\text{Level 1}} + \text{Level Sums}^2_{\text{Level 2}}}{n} \right) - \left[\frac{(\Sigma\, x_i)^2}{N} \right] \qquad (9.1)$$

where $\quad SS_x$ = sums of squares for factor x

$\text{Level Sum}_{\text{Level 1}}$ = level sum for factor x at level 1

$\text{Level Sum}_{\text{Level 2}}$ = level sum for factor x at level 2

n_x = number of data points used in calculating the level

sums for either level 1 or level 2

N = total number of data points in the experiment

Table 9.8. Secondary Check for Errors in the Level Sums Table

Factor	Level 1	Level 2	Total Level 1 + Level 2
Bat weight	4.545	5.840	10.385
Bat length	4.107	6.278	10.385
Bat material	5.565	4.820	10.385
Distance to pitcher	4.994	5.391	10.385
Distance from plate	5.422	4.963	10.385
Footwear	4.470	5.915	10.385
Stance	5.512	4.873	10.385
Totals	34.615 + 38.08	= 72.695	72.695

The next calculation that must be done is to obtain the total variation in the system or SS_T. Equation (9.2) describes this parameter.

$$SS_T = \sum_{i=1}^{N} x_i^2 - \left[\frac{(\Sigma x_i)^2}{N} \right] \tag{9.2}$$

where

SS_T = total variation in the system

$\displaystyle\sum_{i=1}^{N} x^2$ = the sum of each individual data point squared

$\left[\dfrac{(\Sigma x_i)^2}{N} \right]$ = correction factor

Table 9.9 summarizes the sums of squares calculations for each factor and the total variation, SS_T, in this experiment.

Alternately, the total sums of squares, SS_T, can also be calculated by summing the individual sums of squares calculations as shown in Eq. (9.3).

$$SS_T = SS_A + SS_B + \cdots + SS_F + SS_G \tag{9.3}$$

9.5 Significance Plot

The purpose of a screening experiment is to identify the significant few from the trivial many. How does an experimenter start deciding which factors are worth further investigation? There are many ways to do this, but one of the simplest is to make a significance plot. This plot is similar to a Pareto diagram. The SS_x for each factor are plotted in descending order of magnitude from the left to the right on a

Table 9.9. Sums of Squares Calculations

$$SS_{BatWt} = \left(\frac{4.545^2 + 5.840^2}{16} \right) - \frac{10.382^2}{32} = 3.423 - 3.370 = 0.053$$

$$SS_{BatLth} = \left(\frac{4.107^2 + 6.278^2}{16} \right) - \frac{10.385^2}{32} = 3.518 - 3.370 = 0.148$$

$$SS_{BatMtl} = \left(\frac{5.565^2 + 4.820^2}{16} \right) - \frac{10.385^2}{32} = 3.388 - 3.370 = 0.018$$

$$SS_{DisPit} = \left(\frac{4.994^2 + 5.391^2}{16} \right) - \frac{10.385^2}{32} = 3.375 - 3.370 = 0.005$$

$$SS_{DisPlt} = \left(\frac{5.422^2 + 4.963^2}{16} \right) - \frac{10.385^2}{32} = 3.377 - 3.370 = 0.007$$

$$SS_{Footwear} = \left(\frac{4.470^2 + 5.915^2}{16} \right) - \frac{10.385^2}{32} = 3.436 - 3.370 = 0.066$$

$$SS_{Stance} = \left(\frac{5.512^2 + 4.873^2}{16} \right) - \frac{10.385^2}{32} = 3.383 - 3.370 = 0.013$$

$$SS_T = (.210^2 + .225^2 + .210^2 + .220^2 + \ldots + .420^2 + .413^2 + .428^2 + .410^2) - 3.370 = 0.335$$

piece of normal graph paper. However, instead of a bar chart each point on the graph is connected by a solid line.

It is usually quite obvious from the graph which factors have had the greatest effect on the quality characteristic (i.e., the dependent factor) and which have not. The factors along the steepest section of the graph are the more important ones and those along the flat portion or the bottom of the slope are the least important. The significance plot for this investigation is shown in Figure 9.2. From this plot, factor B would be considered the most significant factor followed by factors F and A. All remaining factors, C, G, E, and D, would not be considered significant in this experiment.

9.6 Pooled Error ANOVA Table

In an experimental array where all columns have been filled (i.e., referred to as a saturated design), no allowances for measuring the experimental error have been made. Since the premise of an ANOVA calculation is to compare the contribution made by each factor to the explained variation to that of the unexplained variation (i.e., experimental error or noise), a measure of the experimental error is essential. Therefore, an alternate approach must be taken to obtain an estimate of this parameter.

What is commonly done is to group the factors in the experiment that have been shown to have little or no effect on the response variables. Factors that would be grouped together would be those calculated to have the smallest sums of squares. Grouping of factors in this manner is referred to as pooling. The parameter that results from this grouping is appropriately called the pooled error term and is represented by the symbol ep, or simply error.

Looking at the significance plot for this example, factors C, G, E, and D would be considered the least significant factors in this experiment and would logically

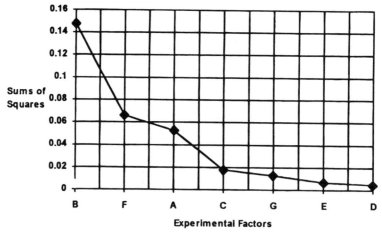

Figure 9.2. Significance plot.

Table 9.10. Adjusted Sums of Squares Data

Factor	SS_x
Bat length	0.148
Footwear	0.066
Bat weight	0.053
ep	0.043
Total	0.335

be grouped together into the pooled error term. These results are summarized in Table 9.10.

9.7 Variance

To be able to calculate the variance for each of the critical factors identified in Table 9.10, the degrees of freedom, df_x, must be calculated [e.g., Eq. (7.5)]. Variance is then determined by dividing the sums of squares for each parameter by the degrees of freedom as shown in Table 9.11.

The degrees of freedom for the pooled error term equal four in this example because the degrees of freedom, in mathematical terms, have additive properties. Each factor combined to form the error term had one degree of freedom associated with it. Since there were four terms added together and each parameter had one degree of freedom, the resultant value for the pooled error term will be four.

9.8 Calculation of the F Statistic

This information can now be transferred to an ANOVA table as shown in Table 9.12, which includes the calculation of the theoretical and experimental F statistic.

Table 9.11. Variance Calculations for Batting Experiment

Factor	$df_x \ (n_x - 1)$	SS_x	$V_x \left(\dfrac{SS_x}{df_x} \right)$
Bat length	1	0.148	0.148
Footwear	1	0.066	0.066
Bat weight	1	0.053	0.053
ep	4	0.043	0.011
Total	7	0.335	0.048

Table 9.12. Calculation of the F Statistic

Factor	$df_x\ (n-1)$	SS_x	$V_x\left(\dfrac{SS_x}{df_x}\right)$	$F\left(\dfrac{V_x}{V_{ep}}\right)$	F_{Table} $F_{(1,\,4)0.05}$
Bat length	$(2-1)=1$	0.148	$\left(\dfrac{0.148}{1}\right)=0.148$	$\left(\dfrac{0.148}{0.011}\right)=13.46$	
Footwear	$(2-1)=1$	0.066	$\left(\dfrac{0.066}{1}\right)=0.066$	$\left(\dfrac{0.066}{0.011}\right)=6.00$	
Bat weight	$(2-1)=1$	0.053	$\left(\dfrac{0.053}{1}\right)=0.053$	$\left(\dfrac{0.053}{0.011}\right)=4.82$	7.75
ep	4	0.043			
Total	$(8-1)=7$	0.335			

9.9 Expected Sums of Squares and Percent Contribution

SS'_x or the expected sums of squares, and P, the percent contribution, are two additional parameters that can be calculated and incorporated into an ANOVA table. The expected sums of squares is calculated to compensate for any experimental error that may have influenced the calculation of the sum of squares. The percent contribution, P, is used to estimate the proportion of the variation that can be attributed to a specific factor in the experiment.

To calculate these additional parameters Eqs. (9.4) and (9.5) are used for the percent contribution and expected sums of squares, respectively.

$$P = \left(\frac{SS'_x}{SS_T}\right) \times 100 \tag{9.4}$$

$$SS'_x = SS_x - (V_{ep} \times df_x) \tag{9.5}$$

where $\quad SS_T$ = total sums of squares

$\qquad\quad SS_x$ = sums of squares for factor x

$\qquad\quad V_{ep}$ = variance of the pooled error term

$\qquad\quad df_x$ = degrees of freedom associated with factor x

Calculation of these parameters is shown in Table 9.13.

The percent contribution due to error, P, is an important factor as it offers a quantitative evaluation of the experimental results. If the sum of the percent contribution for each factor in the ANOVA table does not total approximately 70–75%, the data are indicating that additional investigation may be warranted. There is a possibility that an important factor has been missed, experimental conditions were not controlled adequately, or measurement of the response may have been flawed.

There will also be situations where a factor may be determined to be statistically insignificant according to the F statistic, but that it still has a sizable percent

Table 9.13. Calculation of SS'_x and P

Factor	df_x	SS_x	SS'_x	P
Bat length	1	0.148	$0.148 - (0.011 \times 1) = 0.137$	$\left(\dfrac{0.137}{0.335}\right) \times 100 = 40.9$
Footwear	1	0.066	$0.066 - (0.011 \times 1) = 0.055$	$\left(\dfrac{0.055}{0.335}\right) \times 100 = 16.4$
Bat weight	1	0.053	$0.053 - (0.011 \times 1) = 0.042$	$\left(\dfrac{0.042}{0.335}\right) \times 100 = 12.5$
ep	4	0.043		
Total	7	0.035		

contribution. If this situation occurs this factor should be considered to have an influence on the response variable and continued investigation is warranted. Any factor showing a percent contribution of greater than 10% may be of interest and should not be discounted without additional experimentation to determine its importance.

The completed Pooled Error ANOVA table, combining all the above calculations, is illustrated in Table 9.14.

9.10 Interpretation of ANOVA Table

Comparison of the calculated F statistic to that obtained from the F table suggests that only the bat length is statistically significant. However, according to the percent contribution both the bat weight and the use of spikes have an effect on the desired outcome in this experiment. From a practical point of view it is safer to think a factor is significant and receive only minor improvements than to eliminate this factor from additional study. If a factor is eliminated from the research program it may be impossible to optimize the process because this factor was eliminated early in the process. This, of course, will be true only if significant costs or increase in labor are not being incurred as a result of controlling a factor that is marginally significant.

Table 9.14. Completed Pooled Error ANOVA Table

Factor	df_x	SS_x	$V_x \left(\dfrac{SS_x}{df_x}\right)$	$F \left(\dfrac{V_x}{V_{ep}}\right)$	F_{Table}	SS'_x	P
Bat length	1	0.148	0.148	13.46	7.71	0.137	41
Footwear	1	0.066	0.066	6.00	7.71	0.055	16
Bat weight	1	0.053	0.053	4.82	7.71	0.042	13
ep	4	0.043	0.011				
Total	7	0.335	0.044				

Using this reasoning all three factors, bat length, bat weight, and the use of spikes, have an effect on the baseball player's batting average and should be included in further studies. There is also approximately 30% of the percent contribution that cannot be attributed to any of the factors studied in this investigation. This suggests that there may be other factors or possibly an interaction, not yet identified, that could also influence the player's batting average. Based on these data, the batting coach should continue to look for other factors that may be affecting the ball player's batting average before ending the search for important factors.

9.11 Estimation of Best Run Conditions

After completing the above data analysis a question likely to be foremost on the mind of the batting coach is "What can be expected if the player uses the conditions suggested by the experiment?" If a full factorial experiment had been run (i.e., only three variables studied in the L_8 array instead of seven) it could be possible that the optimal combination of process factors was run during the experiment. In this instance, what would be required to obtain an estimation of the expected results would be to use the average obtained for the trial containing the recommended factor levels.

Estimating the mean response in a saturated orthogonal array, Eq. (9.6) is used. This equation is used because a linear relationship among the process variables has been assumed.

$$\hat{\mu} = \bar{T} + (\overline{LS}_{x1} - \bar{T}) + (\overline{LS}_{x2} - \bar{T}) + \cdots + (\overline{LS}_{xn} - \bar{T}) \tag{9.6}$$

where $\hat{\mu}$ = estimate of the mean response

$\bar{T}$ = mean of all experimental data

$= \dfrac{\Sigma x}{N} = \dfrac{sum \text{ of all data points}}{\text{number of data points in experiment}}$

$\overline{LS}_{xn}$ = optimal level sum response for the significant factor

at the level of interest

The key factors for the baseball player example and their respective level sum responses are summarized in Table 9.15. Using these values for the significant factors the predicted mean response for this example is as follows:

Table 9.15. Parameters for Calculation of Predicted Results

Factor	Most Significant Level	Level Sum Response
Bat length	Level 2	.392
Footwear	Level 2	.369
Bat weight	Level 2	365
Total		10.385

$$\hat{\mu} = \frac{10.385}{32} + \left(.365 - \frac{10.385}{32}\right) + \left(.392 - \frac{10.385}{32}\right) + \left(.369 - \frac{10.385}{32}\right)$$

$$= (.365 + .392 + .369) + \left(\frac{10.385}{32} - \frac{10.385}{32} - \frac{10.385}{32} - \frac{10.3856}{32}\right)$$

$$= 1.126 - 2\left(\frac{10.385}{32}\right)$$

$$= 1.126 - .659$$

$$= .477$$

Based on the above prediction, the slumping baseball player could be expected to hit for a 0.477 average by changing his batting technique to include a long and heavy bat, and spikes.

9.12 Confidence Intervals

Estimation of the mean response is meaningful only if there is also some idea of the spread that can be expected in the data. The importance of having knowledge of the spread in the data was demonstrated by the blood alcohol experiment discussed in Chapter 6. An estimation of the spread in the data can be obtained by calculating a confidence interval or CI.

The error of the estimate is defined by Eq. (9.7):

$$\text{Error} = \sqrt{\left(\frac{F_{\text{df}_x;\ \text{df}_{ep}} \times V_{ep}}{n_{\text{eff}}}\right)} \qquad (9.7)$$

where $F_{\text{df}_x;\ \text{df}_{ep}}$ = F statistic associated with the specified α-risk

and the degrees of freedom for each factor

in the experiment, df_x, and the degrees of error term, df_{ep}

V_{ep} = variance for the pooled error term

n_{eff} = effective number of df for the error

$$= \frac{N}{1 + (\textit{total number of } \text{df}_x \textit{ of the factors used to calculate } \hat{\mu})}$$

N = number of trials in the array used for the experiment

Calculating the error for this example

$$df_x = 1 \qquad df_{ep} = 4 \therefore \qquad F_{df_x; \, df_{ep}} = 7.75$$

$$V_{ep} = 0.011 \qquad n_{eff} = \left(\frac{8}{1 + (1 + 1 + 1)} = 2 \right)$$

Using this information the error for this estimate is

$$\text{error} = \sqrt{\left(\frac{7.75 \times 0.011}{2} \right)} = \sqrt{0.0426}$$

$$= \pm .206$$

Therefore the confidence interval would be stated as follows:

$$CI = 0.477 \pm 0.206$$

This means that the slumping baseball player would expect to see his batting average fall between 0.271 and 0.683 if he changed from his current bat to a longer and heavier one and if his footwear had spikes.

9.13 Confirmation Experiment

What needs to be done to complete the experimental program is to determine whether the predicted batting average will be achieved under actual playing conditions. This can be accomplished by conducting what is commonly referred to as a confirmation experiment. A confirmation experiment consists of adopting the recommended levels of the key factors (i.e., a heavy, large bat and spiked shoes) and the most favorable settings (i.e., economically or ease of use) of all remaining factors investigated in the experiment. The batting average of the player would then be monitored over a finite period of time to determine the effect of the changes. Predictions would be considered confirmed if the player's batting average fell within the calculated interval. If this occurred the changes would be adopted on a permanent basis.

If, however, the player's batting average was not near the predicted mean, it suggests that there are other factors that are also important, the measurement system might not have been a good indicator of the problem, or possibly an interaction may exist among the factors studied. Additional experimentation would be required to obtain a closer approximation of the player's true batting average.

A

Glossary

Action Step: A part of a flow diagram describing the next logical operation in a process.

Alternate Hypothesis (H_a): States there is sufficient evidence to conclude that a statistically significant difference does exist between the two materials; products, processes, etc., being investigated.

Analysis of Variance: A statistical technique in which the magnitude of the change in a measured response is compared to the magnitude of the calculated experimental error rather than being directly compared to a sample mean.

ANOVA: *See* analysis of variance.

ANOVA Table: A tabular template used for the presentation of results from an analysis of variance calculation.

Attribute Data: Discrete data points that are generated as a result of performing a particular operation.

Average: An indicator of where most of the data in a sample are clustered. It is also called the mathematical center of the data.

Balanced Designs: Each factor in the array will be compared to all other factors in a design an equal number of times.

Between-Data Variation: Difference between two independent sets of data points.

Cause and Effect Diagram: An illustration designed to show how a variable/factor, called the cause, relates to a specific outcome, called the effect.

Check Sheet: A template used to record the frequency of occurrence of an item or items in a process.

Confidence Intervals: The limits of the interval around the sample statistic that is believed to include the population parameter.

Confounding: The experimental situation in which main effects have been mingled with other main effects or multifactor interactions. This means that there are factor combinations associated with each column in an array other than the main factor that has been directly assigned to a particular column by the experimenter.

Controllable Factor: *See* indepenent factor.

Correction Factor: Mathematical estimation, used in the analysis of variance, of the total magnitude of the variation caused by all the variables being studied in an experiment.

Cross Product: Representative of the interaction between two factors. It is obtained by multiplying the signs of the coded factors in an array together and is denoted by a $\times$ b. The cross product of positive and negative signs will result in a negative sign and the cross product of two positive or two negative signs culminates in a positive sign.

Customer: Any company or individual for whom a person provides a product or service. This includes individuals within one's own company as well as the traditional customer who buys the finished product or service.

Decision Step: A point in a flow diagram where a question must be answered before being able to continue along the main flow path of the diagram.

Degrees of Freedom: The number of independent parameters associated with an experiment or factor. If a particular factor or experiment has a large number of degrees of freedom associated with it, a greater amount of statistical information can be obtained about the factor or experiment.

Denominator: The variable or equation that is found in the bottom half of a fraction or mathematical formula.

Experiment: The general term used to describe the manipulation of one factor or group of factors to observe what effect they have had on a particular process or product.

Expected Sums of Squares: A corrected estimate (an error component is subtracted) of a factor's contribution to the observed variation.

Experimental Error: A function of the various uncontrollable factors that contribute to variations in a measurement.

External Customer: Anyone who purchases a product or a service from your organization who is not directly associated with your company.

F Statistic: The ratio of the measured variances for the different treatments in an experiment. The largest variance is always found in the numerator of the equation and the smallest is located in the denominator. If the calculated experimental F statistic is larger than a theoretical determined one, the factor is considered to be statistically significant. If the converse is true, the factor is considered to be statistically insignificant.

Flow Diagram: A graphic representation of the logical flow through a "process."

Group 1 Factors: Variables that are quite difficult to change or take a long time to show the effects of a change (e.g., temperature or materials).

Group 2 Factors: Factors that are easy to change or for which the effects of change are observed immediately (e.g., pressure or time).

Independent Factors: A process or product parameter that can be changed from one setting to another.

Interaction: The relationship between two factors whereby, as a result of changing one factor, the response of the other factor is also affected.

Internal Customer: Anybody within your organization for whom you could perform a service or produce a product.

Level: A particular setting used for either a controllable or noncontrollable factor in an experiment. A level or setting is typically denoted in Taguchi methodology by consecutive whole numbers such as 1, 2, 3, 4, . . . , n. In classical designs a $+1$ and -1 sign are commonly used to denote levels for a factor.

Level Sums: An estimation of each individual factor's contribution to the overall variation observed in the experiment.

Noncontrollable Factor: A factor that cannot easily be adjusted (or adjusted at all) or would be highly undesirable to control in a given manufacturing process or product, for example, environmental factors such as humidity, lot-to-lot variations in raw materials, and time of day.

Normal Distribution: A continuous bell-shaped frequency distribution generated from process data where the data are affected only by common causes of variation.

Null Hypothesis (H_0): States that any differences in observed output between two process treatments are statistically insignificant and therefore cannot be attributed to the change made in a "*process*" but is more likely the result of random chance or experimental error.

Numerator: Top half of a fraction or mathematical formula.

Orthogonal Array: A set of equations in which the sum of the coefficients found in the equations that describe a particular experimental array (also called contrasts) will be equal to zero. This means the experiment is balanced where each factor is compared to another an equal number of times at both their high and low levels.

Pareto Principle: 80 percent of the opportunities results from 20 percent of problems. In other words, most effects come from only a few causes.

Plackett–Burman Designs: Statistical designs used for screening experiments; similar to orthogonal arrays.

Percent Contribution, P: Used to estimate the proportion of the variation attributable to a specific factor in an experiment.

Population: The total number of observations that could be made about a particular process or product.

Process: The combination of people, raw materials, equipment, and procedures used to make a product or provide a service. This definition is applicable to any type of activity (e.g., delivering mail or making a widgit).

Quality Characteristic: A parameter used as an absolute measure of the performance of the process or product (e.g., pressure, tensile strength, or elongation). It is the same as a dependent parameter.

Response Factor: An absolute measure of the performance of the process or product; equivalent to a quality characteristic.

Saturated Fractional Factorial Experiment: A type of experimental design that does not confound main effects with other main effects but does confound two factor interactions with main effects since all columns in the design are filled with a process factor.

Screening Designs: A series of experimental designs used to investigate a large number of factors in the minimal number of trials. Every column in a screening design is filled with a process factor.

Secondary Loop (Decision Loop): Used to describe to the experimenter the steps required once the main flow path has been left.

Significance Plot: A graphic representation of the sum of squares of each factor investigated in an experiment. This plot is similar to a Pareto diagram as the sum of squares for each factor is plotted in descending order of magnitude from the left to the right on normal graph paper. However, instead of a bar chart each point on the graph is connected by a solid line. The factors located on the steepest portion of the plot are considered most significant in an experiment.

Standard Deviation: A calculated measure of variation within a group of data; it reflects the average distance between a particular point and the overall average.

Sum of Squares: The numerator of the variance formula.

Table of Reasonableness: A special type of table that lists each trial combination in the experiment once the factor ranges have been chosen. This allows for a visual review of the various process conditions prior to running them in production.

Total Variation: The magnitude of the spread or differences observed between all the data in a given data set.

Variable: Something that is associated with and can be used to influence the operation or quality of a particular process, product, or event. Factor is another commonly used term for variable.

Variable Data: Quantitative data obtained from measurement of a given parameter such as length, temperature, or concentration.

Variance: A measure of the spread or dispersion of data; it is the square of the standard deviation.

Variation: The unavoidable differences between results obtained from a process or measurement for whatever reason.

Within-Data Variation: Scatter observed within a given set of data.

B

Common Two- and Three-Level Orthogonal Arrays[1]

L_4

	Factors		
Trials	A	B	C
1	1	1	1
2	1	2	2
3	2	1	2
4	2	2	1

Confounding Assignments—Full Factorial		
A	B	A × B

---

[1] From "Orthogonal Arrays and Linear Graphs." American Supplier Institute, Inc., 1986. Reprinted with permission.

L_8

	Factors						
Trials	1	2	3	4	5	6	7
1	1	1	1	1	1	1	1
2	1	1	1	2	2	2	2
3	1	2	2	1	1	2	2
4	1	2	2	2	2	1	1
5	2	1	2	1	2	1	2
6	2	1	2	2	1	2	1
7	2	2	1	1	2	2	1
8	2	2	1	2	1	1	2

	Confounding Assignments—Full Factorial						
	A	B	A × B	C	A × C	B × C	A × B × C

L_9

	Factors			
Trials	1	2	3	4
1	1	1	1	1
2	1	2	2	2
3	1	3	3	3
4	2	1	2	3
5	2	2	3	1
6	2	3	1	2
7	3	1	3	2
8	3	2	1	3
9	3	3	2	1

	Confounding Assignments—Full Factorial		
	A	B	A × B	A × B

L_{12}

Trials						Factors					
	1	2	3	4	5	6	7	8	9	10	11
1	1	1	1	1	1	1	1	1	1	1	1
2	1	1	1	1	1	2	2	2	2	2	2
3	1	1	2	2	2	1	1	1	2	2	2
4	1	2	1	2	2	1	2	2	1	1	2
5	1	2	2	1	2	2	1	2	1	2	1
6	1	2	2	2	1	2	2	1	2	1	1
7	2	1	2	2	1	1	2	2	1	2	1
8	2	1	2	1	2	2	2	1	1	1	2
9	2	1	1	2	2	2	1	2	2	1	1
10	2	2	2	1	1	1	1	2	2	1	2
11	2	2	1	2	1	2	1	1	1	2	2
12	2	2	1	1	2	1	2	1	2	2	1

No Confounding Patterns Associated with This Design

L_{16}

Trials								Factors							
	1	2	3	4	5	6	7	8	9	10	11	12	13	14	15
1	1	1	1	1	1	1	1	1	1	1	1	1	1	1	1
2	1	1	1	1	1	1	1	2	2	2	2	2	2	2	2
3	1	1	1	2	2	2	2	1	1	1	1	2	2	2	2
4	1	1	1	2	2	2	2	2	2	2	2	1	1	1	1
5	1	2	2	1	1	2	2	1	1	2	2	1	1	2	2
6	1	2	2	1	1	2	2	2	2	1	1	2	2	1	1
7	1	2	2	2	2	1	1	1	1	2	2	2	2	1	1
8	1	2	2	2	2	1	1	2	2	1	1	1	1	2	2
9	2	1	2	1	2	1	2	1	2	1	2	1	2	1	2
10	2	1	2	1	2	1	2	2	1	2	1	2	1	2	1
11	2	1	2	2	1	2	1	1	2	1	2	2	1	2	1
12	2	1	2	2	1	2	1	2	1	2	1	1	2	1	2
13	2	2	1	1	2	2	1	1	2	2	1	1	2	2	1
14	2	2	1	1	2	2	1	2	1	1	2	2	1	1	2
15	2	2	1	2	1	1	2	1	2	2	1	2	1	1	2
16	2	2	1	2	1	1	2	2	1	1	2	1	2	2	1

Confounding Assignments—Full Factorial

1	2	3	4	5	6	7	8	9	10	11	12	13	14	15
A	B	A	C	A	B	A	D	A	B	A	C	A	B	A
		×		×	×	×		×	×	×	×	×	×	×
		B		C	C	B		D	D	B	D	C	C	B
						×				×		×	×	×
						C				D		D	D	C
														×
														D

L_{18}

Trials	Factors							
	1	2	3	4	5	6	7	8
1	1	1	1	1	1	1	1	1
2	1	1	2	2	2	2	2	2
3	1	1	3	3	3	3	3	3
4	1	2	1	1	2	2	3	3
5	1	2	2	2	3	3	1	1
6	1	2	3	3	1	1	2	2
7	1	3	1	2	1	3	2	3
8	1	3	2	3	2	1	3	1
9	1	3	3	1	3	2	1	2
10	2	1	1	3	3	2	2	1
11	2	1	2	1	1	3	3	2
12	2	1	3	2	2	1	1	3
13	2	2	1	2	3	1	3	2
14	2	2	2	3	1	2	1	3
15	2	2	3	1	2	3	2	1
16	2	3	1	3	2	3	1	2
17	2	3	2	1	3	1	2	3
18	2	3	3	2	1	2	3	1

No Confounding Patterns Associated with This Design

L_{27}

Trials	Factors												
	1	2	3	4	5	6	7	8	9	10	11	12	13
1	1	1	1	1	1	1	1	1	1	1	1	1	1
2	1	1	1	1	2	2	2	2	2	2	2	2	2
3	1	1	1	1	3	3	3	3	3	3	3	3	3
4	1	2	2	2	1	1	1	2	2	2	3	3	3
5	1	2	2	2	2	2	2	3	3	3	1	1	1
6	1	2	2	2	3	3	3	1	1	1	2	2	2
7	1	3	3	3	1	1	1	3	3	3	2	2	2
8	1	3	3	3	2	2	2	1	1	1	3	3	3
9	1	3	3	3	3	3	3	2	2	2	1	1	1

(Continued)

L_{27} (*Continued*)

Trials						Factors							
	1	2	3	4	5	6	7	8	9	10	11	12	13
10	2	1	2	3	1	2	3	1	2	3	1	2	3
11	2	1	2	3	2	3	1	2	3	1	2	3	1
12	2	1	2	3	3	1	2	3	1	2	3	1	2
13	2	2	3	1	1	2	3	2	3	1	3	1	2
14	2	2	3	1	2	3	1	3	1	2	1	2	3
15	2	2	3	1	3	1	2	1	2	3	2	3	1
16	2	3	1	2	1	2	3	3	1	2	2	3	1
17	2	3	1	2	2	3	1	1	2	3	3	1	2
18	2	3	1	2	3	1	2	2	3	1	1	2	3
19	3	1	3	2	1	3	2	1	3	2	1	3	2
20	3	1	3	2	2	1	3	2	1	3	2	1	3
21	3	1	3	2	3	2	1	3	2	1	3	2	1
22	3	2	1	3	1	3	2	2	1	3	3	2	1
23	3	2	1	3	2	1	3	3	2	1	1	3	2
24	3	2	1	3	3	2	1	1	3	2	2	1	3
25	3	3	2	1	1	3	2	3	2	1	2	1	3
26	3	3	2	1	2	1	3	1	3	2	3	2	1
27	3	3	2	1	3	2	1	2	1	3	1	3	2

Confounding Assignments—Full Factorial

	1	2	3	4	5	6	7	8	9	10	11	12	13
	A	B	A × B	A × B²	C	A × C	A × C²	B × C	A × B × C	A × B² × C²	B × C²	A × B² × C	A × B × C²

L_{32}

Trials	Factors																														
	1	2	3	4	5	6	7	8	9	10	11	12	13	14	15	16	17	18	19	20	21	22	23	24	25	26	27	28	29	30	31
1	1	1	1	1	1	1	1	1	1	1	1	1	1	1	1	1	1	1	1	1	1	1	1	1	1	1	1	1	1	1	1
2	1	1	1	1	1	1	1	1	1	1	1	1	1	1	1	2	2	2	2	2	2	2	2	2	2	2	2	2	2	2	2
3	1	1	1	1	1	1	1	2	2	2	2	2	2	2	2	1	1	1	1	1	1	1	1	2	2	2	2	2	2	2	2
4	1	1	1	1	1	1	1	2	2	2	2	2	2	2	2	2	2	2	2	2	2	2	2	1	1	1	1	1	1	1	1
5	1	1	1	2	2	2	2	1	1	1	1	2	2	2	2	1	1	1	1	2	2	2	2	1	1	1	1	2	2	2	2
6	1	1	1	2	2	2	2	1	1	1	1	2	2	2	2	2	2	2	2	1	1	1	1	2	2	2	2	1	1	1	1
7	1	1	1	2	2	2	2	2	2	2	2	1	1	1	1	1	1	1	1	2	2	2	2	2	2	2	2	1	1	1	1
8	1	1	1	2	2	2	2	2	2	2	2	1	1	1	1	2	2	2	2	1	1	1	1	1	1	1	1	2	2	2	2
9	1	2	2	1	1	2	2	1	1	2	2	1	1	2	2	1	1	2	2	1	1	2	2	1	1	2	2	1	1	2	2
10	1	2	2	1	1	2	2	1	1	2	2	1	1	2	2	2	2	1	1	2	2	1	1	2	2	1	1	2	2	1	1
11	1	2	2	1	1	2	2	2	2	1	1	2	2	1	1	1	1	2	2	1	1	2	2	2	2	1	1	2	2	1	1
12	1	2	2	1	1	2	2	2	2	1	1	2	2	1	1	2	2	1	1	2	2	1	1	1	1	2	2	1	1	2	2
13	1	2	2	2	2	1	1	1	1	2	2	2	2	1	1	1	1	2	2	2	2	1	1	1	1	2	2	2	2	1	1
14	1	2	2	2	2	1	1	1	1	2	2	2	2	1	1	2	2	1	1	1	1	2	2	2	2	1	1	1	1	2	2
15	1	2	2	2	2	1	1	2	2	1	1	1	1	2	2	1	1	2	2	2	2	1	1	2	2	1	1	1	1	2	2
16	1	2	2	2	2	1	1	2	2	1	1	1	1	2	2	2	2	1	1	1	1	2	2	1	1	2	2	2	2	1	1
17	2	1	2	1	2	1	2	1	2	1	2	1	2	1	2	1	2	1	2	1	2	1	2	1	2	1	2	1	2	1	2
18	2	1	2	1	2	1	2	1	2	1	2	1	2	1	2	2	1	2	1	2	1	2	1	2	1	2	1	2	1	2	1
19	2	1	2	1	2	1	2	2	1	2	1	2	1	2	1	1	2	1	2	1	2	1	2	2	1	2	1	2	1	2	1
20	2	1	2	1	2	1	2	2	1	2	1	2	1	2	1	2	1	2	1	2	1	2	1	1	2	1	2	1	2	1	2
21	2	1	2	2	1	2	1	1	2	1	2	2	1	2	1	1	2	1	2	2	1	2	1	1	2	1	2	2	1	2	1
22	2	1	2	2	1	2	1	1	2	1	2	2	1	2	1	2	1	2	1	1	2	1	2	2	1	2	1	1	2	1	2
23	2	1	2	2	1	2	1	2	1	2	1	1	2	1	2	1	2	1	2	2	1	2	1	2	1	2	1	1	2	1	2
24	2	1	2	2	1	2	1	2	1	2	1	1	2	1	2	2	1	2	1	1	2	1	2	1	2	1	2	2	1	2	1
25	2	2	1	1	2	2	1	1	2	2	1	1	2	2	1	1	2	2	1	1	2	2	1	1	2	2	1	1	2	2	1
26	2	2	1	1	2	2	1	1	2	2	1	1	2	2	1	2	1	1	2	2	1	1	2	2	1	1	2	2	1	1	2

(Continued)

L₃₂ Continued.

Factors

Trials	1	2	3	4	5	6	7	8	9	10	11	12	13	14	15	16	17	18	19	20	21	22	23	24	25	26	27	28	29	30	31
27	2	2	1	1	2	2	1	2	1	1	2	2	1	1	2	1	2	2	1	1	2	2	1	2	1	1	2	2	1	1	2
28	2	2	1	1	2	2	1	2	1	1	2	2	1	1	2	2	1	1	2	2	1	1	2	1	2	2	1	1	2	2	1
29	2	2	1	2	1	1	2	1	2	2	1	2	1	1	2	1	2	2	1	2	1	1	2	1	2	2	1	2	1	1	2
30	2	2	1	2	1	1	2	1	2	2	1	2	1	1	2	2	1	1	2	1	2	2	1	2	1	1	2	1	2	2	1
31	2	2	1	2	1	1	2	2	1	1	2	1	2	2	1	1	2	2	1	2	1	1	2	2	1	1	2	1	2	2	1
32	2	2	1	2	1	1	2	2	1	1	2	1	2	2	1	2	1	1	2	1	2	2	1	1	2	2	1	2	1	1	2

Confounding Assignments—Full Factorial

1	2	3	4	5	6	7	8	9	10	11	12	13	14	15	16	17	18	19	20	21	22	23	24	25	26	27	28	29	30	31
A	B	A×B	C	A×C	B×C	A×B×C	D	A×D	B×D	A×B×D	C×D	A×C×D	B×C×D	A×B×C×D	E	A×E	B×E	A×B×E	C×E	A×C×E	B×C×E	A×B×C×E	D×E	A×D×E	B×D×E	A×B×D×E	C×D×E	A×C×D×E	B×C×D×E	A×B×C×D×E

C

Confounding Patterns for Common Orthogonal Arrays[1]

L_8

For 4 factors in 8 trials all main factors are clear of 2-factor interactions, but 2-factor interactions are confused with one another. The aliases for this design are

$\quad\quad\quad A = B \times C \times D$

$\quad\quad\quad B = A \times C \times D$

$\quad\quad\quad C = A \times B \times D$

$\quad\quad\quad D = A \times B \times C$

[1] Confounding patterns reprinted with permission from the Design-Ease® statistical design of experiment software from Stat-Ease, Minneapolis, MN.

$$A \times B = C \times D$$

$$A \times C = B \times D$$

$$A \times D = B \times C$$

For 5 factors in 8 trials main factors are clear of other main effects, but main factors and 2-factor interactions are confounded. The aliases for this design are

$$A = B \times D = C \times E$$

$$B = A \times D = C \times D \times E$$

$$C = A \times E = B \times D \times E$$

$$D = A \times B = B \times C \times E$$

$$E = A \times C = B \times C \times D$$

$$B \times C = D \times E = A \times C \times D = A \times B \times E$$

$$B \times E = C \times D = A \times B \times C = A \times D \times E$$

For 6 factors in 8 trials main factors are clear of other main effects, but main factors and 2-factor interactions are confounded. The aliases for this design are

$$A = B \times D = C \times E = C \times D \times F = B \times E \times F$$

$$B = A \times D = C \times F = C \times D \times E = A \times E \times F$$

$$C = A \times E = B \times F = B \times D \times E = A \times D \times F$$

$$D = A \times B = E \times F = B \times C \times E = A \times C \times F$$

$$E = A \times C = D \times F = B \times C \times D = A \times B \times F$$

$$F = B \times C = D \times E = A \times C \times D = A \times B \times E$$

$$A \times F = B \times E = C \times D = A \times B \times C = A \times D \times E = B \times D \times F = C \times E \times F$$

For 7 factors in 8 trials main factors are clear of other main factors, but main factors and 2-factor interactions are confounded. The 2-factor aliases for this design are

$$A = B \times D = C \times E = F \times G$$

$$B = A \times D = C \times F = E \times G$$

$$C = A \times E = B \times F = D \times G$$

$$D = A \times B = E \times F = C \times G$$

$$E = A \times C = D \times F = B \times G$$

$$F = B \times C = D \times E = A \times G$$

$$G = C \times D = B \times E = A \times F$$

L_{16}

For 5 factors in 16 trials main factors are clear of 2-factor interactions and all 2-factor interactions are clear of one another. The aliases for this design are

$$A \times B = C \times D \times E \qquad B \times D = A \times C \times E$$

$$A \times C = B \times D \times E \qquad B \times E = A \times C \times D$$

$$A \times D = B \times C \times E \qquad C \times D = A \times B \times E$$

$$A \times E = B \times C \times D \qquad C \times E = A \times B \times D$$

$$B \times C = A \times D \times E \qquad D \times E = A \times B \times C$$

For 6 factors in 16 trials main factors are clear of 2-factor interactions, but 2-factor interactions are confounded with one another. Tthe aliases for this design are

$$A = B \times C \times E = D \times E \times F \qquad A \times B = C \times E$$

$$B = A \times C \times E = C \times D \times F \qquad A \times C = B \times E$$

$$C = A \times B \times E = B \times D \times F \qquad A \times D = E \times F$$

$$C = B \times C \times F = A \times E \times F \qquad A \times E = B \times C = D \times F$$

$$E = A \times B \times C = A \times D \times F \qquad A \times F = D \times E$$

$$F = B \times C \times D = A \times D \times E \qquad B \times D = C \times F$$

$$B \times F = C \times D$$

$$A \times B \times D = C \times D \times E = A \times C \times F = B \times E \times F$$

$$A \times B \times F = B \times D \times E = A \times C \times D = C \times E \times F$$

For 7 factors in 16 trials all main factors are clear of 2-factor interactions, but 2-factor interactions are confused with one another. The aliases for this design are

$$A = B \times C \times E = D \times E \times F = C \times D \times G = B \times F \times G$$

$$B = A \times C \times E = C \times D \times F = D \times E \times G = A \times F \times G$$

$$C = A \times B \times E = B \times D \times F = A \times D \times G = E \times F \times G$$

$$D = B \times C \times F = A \times E \times F = A \times C \times G = B \times E \times G$$

$$E = A \times B \times C = A \times D \times F = B \times D \times G = C \times F \times G$$

$$F = B \times C \times D = A \times D \times E = A \times B \times G = C \times E \times G$$

$$G = A \times C \times D = B \times D \times E = A \times B \times F = C \times E \times F$$

$$A \times B = C \times E = F \times G \qquad A \times C = B \times E = D \times G$$

A × D = E × F = C × G   A × E = B × C = D × F

A × F = D × E = B × G   A × G = C × D = B × F

B × D = C × F = E × G

A × B × D = C × D × E = A × C × F = B × E × F = B × C × G = A × E × G = D × F × G

For 8 factors in 16 trials all main factors are clear of 2-factor interactions, but 2-factor interactions are confounded with one another. The aliases for this design are

A = C × D × F = B × E × F = B × C × G = D × E × G = B × D × H = C × E × H = F × G × H

B = C × D × E = A × E × F = A × C × G = D × F × G = A × D × H = C × F × H = E × G × H

C = B × D × E = A × D × F = A × B × G = E × F × G = A × E × H = B × F × H = D × G × H

D = B × C × E = A × C × F = A × E × G = B × F × G = A × B × H = E × F × H = C × G × H

E = B × C × D = A × B × F = A × D × G = C × F × G = A × C × H = D × F × H = B × G × H

F = A × C × D = A × B × E = B × D × G = C × G × E = B × C × H = D × E × H = A × G × H

G = A × B × C = A × D × E = B × D × F = C × E × F = C × D × H = B × E × H = A × F × H

H = A × B × D = A × C × E = B × C × F = D × E × F = C × D × G = B × E × G = A × F × G

A × B = E × F = C × G = D × H

A × C = D × F = B × G = E × H

A × D = C × F = E × G = B × H

A × E = B × F = D × G = C × H

A × F = C × D = B × E = G × H

A × G = B × C = D × E = F × H

A × H = B × D = C × E = F × G

For 9 factors in 16 trials main factors are clear of other main factors, but main factors and 2-factor inactions are confounded with one another. The aliases for this design are

A = F × J = B × C × E = D × E × F = C × D × G = B × F × G = B × D × H = C × F × H = E × G × H

B = G × J = A × C × E = C × D × F = D × E × G = A × F × G = A × G × H = E × F × H = C × G × H

C = H × J = A × B × E = B × D × F = A × D × G = E × F × G = D × E × H = A × F × H = B × G × H

D = E × J = B × C × F = A × E × F = A × C × G = B × E × G = A × B × H = C × E × H = F × G × H

E = D × J = A × B × C = A × D × F = B × D × G = C × F × G = C × D × H = B × F × H = A × G × H

F = A × J = B × C × D = A × D × E = A × B × G = C × E × G = A × C × H = B × E × H = D × G × H

G = B × J = A × C × D = B × D × E = A × B × F = C × E × F = B × C × H = A × E × H = D × F × G

H = C × J = A × B × D = C × D × E = A × C × F = B × E × F = B × C × G = A × E × G = D × F × G

J = D × E = A × F = B × G = C × H

A × B = C × E = F × G = D × H = C × D × J = B × F × J = A × G × J = E × H × J

A × C = B × E = D × G = F × H = B × D × J = C × F × J = E × G × J = A × H × J

A × D = E × F = C × G = B × H = B × C × J = A × E × J × D × F × J = G × H × J

A × E = B × C = D × F = G × H = A × D × J = E × F × J = C × G × J = B × H × J

A × G = C × D = B × F = E × H = A × B × J = C × E × J = F × G × J = D × H × J

$A \times H = B \times D = C \times F = E \times G = A \times C \times J = B \times E \times J = D \times G \times J = F \times H \times J$

For 10 factors in 16 trials main factors are clear of other main factors, but main factors and 2-factor interactions are confounded with one another. The 2-factor aliases for this design are

$$A = F \times J = B \times K \qquad B = G \times J = A \times K$$
$$C = H \times J = E \times K \qquad D = E \times J = H \times K$$
$$E = D \times J = C \times K \qquad F = A \times J = G \times K$$
$$G = B \times J = F \times K \qquad H = C \times J = D \times K$$
$$J = D \times E = A \times F \qquad K = A \times B = C \times E$$
$$A \times C = B \times E = D \times G = F \times H$$
$$A \times D = E \times F = C \times G = B \times H$$
$$A \times E = B \times C = D \times F = G \times H$$
$$A \times G = C \times D = B \times F = E \times H$$
$$A \times H = B \times D = C \times F = E \times G$$

For 11 factors in 16 trials main factors are clear of other main factors, but main factors and 2-factor interactions are confounded with one another. The 2-factor aliases for this design are

$$A = F \times J = B \times K = C \times L$$
$$B = G \times J = A \times K = E \times L$$
$$C = H \times J = E \times K = A \times L$$
$$D = E \times J = H \times K = G \times L$$
$$E = D \times J = C \times K = B \times L$$
$$F = A \times J = G \times K = H \times L$$
$$G = B \times J = F \times K = D \times L$$
$$H = C \times J = D \times K = F \times L$$
$$J = D \times E = A \times F = B \times G$$
$$K = A \times B = C \times E = F \times G$$
$$L = A \times C = B \times E = D \times G$$
$$A \times D = E \times F = C \times G = B \times H$$
$$A \times E = B \times C = D \times F = G \times H$$
$$A \times G = C \times D = B \times F = E \times H$$
$$A \times H = B \times D = C \times F - E \times G$$

For 12 factors in 16 trials main factors are clear of other main factors, but main factors and 2-factor interactions are confounded with one another. The 2-factor aliases for this design are

$$A = H \times J = B \times K = C \times L = D \times M$$

$$B = G \times J = A \times K = E \times L = F \times M$$

$$C = F \times J = E \times K = A \times L = G \times M$$

$$D = E \times J = F \times K = G \times L = A \times M$$

$$E = D \times J = C \times K = B \times L = H \times M$$

$$F = C \times J = D \times K = H \times L = B \times M$$

$$G = B \times J = H \times K = D \times L = C \times M$$

$$H = A \times J = G \times K = F \times L = E \times M$$

$$J = D \times E = C \times F = B \times G = A \times H$$

$$K = A \times B = C \times E = D \times F = G \times H$$

$$L = A \times C = B \times E = D \times G = F \times H$$

$$M = A \times D = B \times F = C \times G = E \times H$$

$$A \times E = B \times C = F \times G = D \times H = K \times L = J \times M$$

$$A \times F = B \times D = E \times G = C \times H = J \times L = K \times M$$

$$A \times G = E \times F = C \times D = B \times H = J \times K = L \times M$$

For 13 factors in 16 trials main factors are clear of other main factors, but main factors and 2-factor interactions are confounded with one another. The 2-factor aliases for this design are

$$A = H \times J = B \times K = C \times L = D \times M = E \times N$$

$$B = G \times J = A \times K = E \times L = F \times M = C \times N$$

$$C = F \times J = E \times K = A \times L = G \times M = B \times N$$

$$D = E \times J = F \times K = G \times L = A \times M = H \times N$$

$$E = D \times J = C \times K = B \times L = H \times M = A \times N$$

$$F = C \times J = D \times K = H \times L = B \times M = G \times N$$

$$G = B \times J = H \times K = D \times L = C \times M = F \times N$$

$$H = A \times J = G \times K = F \times L = E \times M = D \times N$$

$$J = D \times E = C \times F = B \times G = A \times H = M \times N$$

$$K = A \times B = C \times E = D \times F = G \times H = L \times N$$

$$L = A \times C = B \times E = D \times G = F \times H = K \times N$$

$$M = A \times D = B \times F = C \times G = E \times H = J \times N$$

$$N = B \times C = A \times E = F \times G = D \times H = K \times L = J \times M$$

$$A \times F = B \times D = E \times G = C \times H = J \times L = K \times M$$

$$A \times G = E \times F = C \times D = B \times H = J \times K = L \times M$$

For 14 factors in 16 trials main factors are clear of other main factors, but main factors and 2-factor interactions are confounded with one another. The 2-factor aliases for this design are

$$A = H \times J = B \times K = C \times L = D \times M = E \times N = F \times O$$

$$B = G \times J = A \times K = E \times L = F \times M = C \times N = D \times O$$

$$C = F \times J = E \times K = A \times L = G \times M = B \times N = H \times O$$

$$D = E \times J = F \times K = G \times L = A \times M = H \times N = B \times O$$

$$E = D \times J = C \times K = B \times L = H \times M = A \times N = G \times O$$

$$F = C \times J = D \times K = H \times L = B \times M = G \times N = A \times O$$

$$G = B \times J = H \times K = D \times L = C \times M = F \times N = E \times O$$

$$H = A \times J = G \times K = F \times L = E \times M = D \times N = C \times O$$

$$J = D \times E = C \times F = B \times G = A \times H = M \times N = L \times O$$

$$K = A \times B = C \times E = D \times F = G \times H = L \times N = M \times O$$

$$L = A \times C = B \times E = D \times G = F \times H = K \times N = J \times O$$

$$M = A \times D = B \times F = C \times G = E \times H = J \times N = K \times O$$

$$N = B \times C = A \times E = F \times G = D \times H = K \times L = J \times M$$

$$O = B \times D = A \times F = E \times G = C \times H = J \times L = K \times M$$

$$A \times G = E \times F = C \times D = B \times H = J \times K = L \times M = N \times O$$

For 15 factors in 16 trials main factors are clear of other main factors, but main factors and 2-factor interactions are confounded with one another. The 2-factor aliases for this design are

$$A = H \times J = B \times K = C \times L = D \times M = E \times N = F \times O = G \times P$$

$$B = G \times J = A \times K = E \times L = F \times M = C \times N = D \times O = H \times P$$

$$C = F \times J = E \times K = A \times L = G \times M = B \times N = H \times O = D \times P$$

$$D = E \times J = F \times K = G \times L = A \times M = H \times N = B \times O = C \times P$$

$$E = D \times J = C \times K = B \times L = H \times M = A \times N = G \times O = F \times P$$

$$F = C \times J = D \times K = H \times L = B \times M = G \times N = A \times O = E \times P$$

$$G = B \times J = H \times K = D \times L = C \times M = F \times N = E \times O = A \times P$$

$$H = A \times J = G \times K = F \times L = E \times M = D \times N = C \times O = B \times P$$

$$J = D \times E = C \times F = B \times G = A \times H = M \times N = L \times O = K \times P$$

$$K = A \times B = C \times E = D \times F = G \times H = L \times N = M \times O = J \times P$$

$$L = A \times C = B \times E = D \times G = F \times H = K \times N = J \times O = M \times P$$

$$M = A \times D = B \times F = C \times G = E \times H = J \times N = K \times O = L \times P$$

$$N = B \times C = A \times E = F \times G = D \times H = K \times L = J \times M = O \times P$$

$$O = B \times D = A \times F = E \times G = C \times H = J \times L = K \times M = N \times P$$

$$P = C \times D = E \times F = A \times G = B \times H = J \times K = L \times M = N \times O$$

L_{32}

For 6 factors in 32 trials main factors are clear of 2-factor interactions and all 2-factor interactions are clear of one another. The aliases for this design are

$$A \times B \times C = D \times E \times F \qquad A \times B \times D = C \times E \times F$$

$$A \times B \times E = C \times D \times F \qquad A \times B \times F = C \times D \times E$$

$$A \times C \times D = B \times E \times F \qquad A \times C \times E = B \times D \times F$$

$$A \times C \times F = B \times D \times E \qquad A \times D \times E = B \times C \times F$$

$$A \times D \times F = B \times C \times E \qquad A \times E \times F = B \times C \times D$$

For 7 factors in 32 trials main factors are clear of 2-factor interactions, but some 2-factor interactions are confounded with one another. The aliases for this design are

$A =$	$A \times B = C \times D \times F = D \times E \times F$	$B \times D = A \times C \times F = A \times E \times F$
$B =$	$A \times C = B \times D \times F$	$B \times E = A \times D \times G$
$C = E \times F \times G$	$A \times D = B \times C \times F = B \times E \times G$	$B \times F = A \times C \times D$
$D =$	$A \times E = B \times D \times G$	$B \times G = A \times D \times E$
$E = C \times F \times G$	$A \times F = B \times C \times D$	$C \times D = A \times B \times F$
$F = C \times E \times G$	$A \times G = D \times D \times E$	$C \times E = F \times G$
$G = C \times E \times F$	$B \times C = A \times D \times F$	$C \times F = A \times B \times D = E \times G$
$C \times G = E \times F$	$A \times C \times E = A \times F \times G$	$C \times D \times E = D \times F \times G$
$D \times E = A \times B \times G$	$A \times C \times G = A \times E \times F$	$C \times D \times G = D \times E \times F$
$D \times F = A \times B \times C$	$B \times C \times E = B \times F \times G$	
$D \times G = A \times B$	$B \times C \times G = B \times E \times F$	

For 8 factors in 32 trials main factors are clear of 2-factor interactions, but some 2-factor interactions are confounded with one another. The aliases for this design are

$A = B \times C \times F = B \times D \times G$	$A \times B = C \times F = D \times G$
$B = A \times C \times F = A \times D \times G$	$A \times C = B \times F = E \times G \times H$

C = A × B × F = D × F × G

D = A × B × G = C × F × G

E =

F = A × B × C = C × D × G

G = A × B × D = C × D × F

H =

B × H = C × D × E = E × F × G

C × E = B × D × H = A × G × H

C × H = B × D × E = A × E × G

D × H = B × C × E = A × E × F

E × G = A × C × H = B × F × H

E × H = B × C × D = A × D × F = A × C × G = B × F × G

F × H = A × D × E = B × E × G

A × C × D = B × D × F = B × C × G = A × F × B

A × B × E = C × E × F = D × E × G

A × B × H = C × F × H = D × G × H

A × D = B × G = E × F × H

A × E = D × F × H = C × G × H

A × F = B × C = D × E × H

A × G = B × D = C × E × H

A × H = D × E × F = C × E × G

B × E = C × D × H = F × G × H

C × D = F × G = B × E × H

C × G = D × F = A × E × H

D × E = B × C × H = A × F × H

E × F = A × D × H = B × G × H

G × H = A × C × E = B × E × F

For 9 factors in 32 trials main factors are clear of 2-factor interactions, but some 2-factor interactions are confounded with one another. The aliases for this design are

A = B × F × G = C × F × H = D × F × G

C = A × F × H = B × G × H = D × H × J

E =

G = A × B × F = B × C × H = B × D × J

J = A × D × F = B × D × G = C × D × H

A × B = F × G = D × E × H = C × E × J

A × D = C × E × G = B × E × H = F × J

A × F = B × G = C × H = D × J

A × H = B × D × E = C × F = E × G × J

B × C = D × E × F = G × H = A × E × J

B × H = A × D × E = C × G = E × F × J

C × D = B × E × F = A × E × G = H × J

B × E = C × D × F = A × D × H = A × C × J = F × H × J

C × E = B × D × F = A × D × G = A × B × J = F × G × J

D × E = B × C × F = A × C × G = A × B × H = F × G × H

E × F = B × C × D = D × G × H = C × G × J = B × H × J

E × G = A × C × D = D × F × H = E × F × J = A × H × J

E × H = A × B × D = D × F × G = B × F × J = A × G × J

E × J = A × B × C = C × F × G = B × F × H = A × G × H

B = A × F × G = C × G × H = D × F × J

D = A × F × J = B × G × J = C × H × J

F = A × B × G = A × C × H = A × D × J

H = A × C × F = B × C × G = C × D × J

A × C = D × E × G = F × H = B × E × J

A × E = C × D × G = B × D × H = B × C × J

        = G × H × J

A × G = C × D × E = B × F = E × H × J

A × J = B × C × E = D × F = E × G × H

B × D = C × E × F = A × E × H = G × J

B × J = A × C × E = D × G = E × F × H

C × J = A × B × E = E × F × G = D × H

A × E × F = B × E × G = C × E × H = D × E × J

For 10 factors in 32 trials main factors are clear of 2-factor interactions, but some 2-factor interactions are confounded with one another. The aliases are

$$A = E \times F \times K = D \times G \times K = C \times H \times K = B \times J \times K$$

$$B = E \times F \times J = D \times G \times J = C \times H \times J = A \times J \times K$$

$$C = E \times F \times H = D \times G \times H = B \times H \times J = A \times H \times K$$

$$D = E \times F \times G = C \times G \times H = B \times G \times J = A \times G \times K$$

$$E = D \times F \times G = C \times F \times H = B \times F \times J = A \times F \times K$$

$$F = D \times E \times G = C \times E \times H = B \times E \times J = A \times E \times K$$

$$G = D \times E \times F = C \times D \times H = B \times D \times J = A \times D \times K$$

$$H = C \times E \times F = C \times D \times G = B \times C \times J = A \times C \times K$$

$$J = B \times E \times F = B \times D \times G = B \times C \times H = A \times B \times K$$

$$K = A \times E \times F = A \times D \times G = A \times C \times H = A \times B \times J$$

$$A \times B = C \times D \times F = C \times E \times G = D \times E \times H = F \times G \times H = J \times K$$

$$A \times C = B \times D \times F = B \times E \times G = D \times E \times J = F \times G \times J = H \times K$$

$$A \times D = B \times C \times F = B \times E \times H = C \times E \times J = F \times H \times J = G \times K$$

$$A \times E = B \times C \times G = B \times D \times H = C \times D \times J = G \times H \times J = F \times K$$

$$A \times F = B \times C \times D = B \times G \times H = C \times G \times J = D \times H \times J = E \times K$$

$$A \times G = B \times C \times E = B \times F \times H = C \times F \times J = E \times H \times J = D \times K$$

$$A \times H = B \times D \times E = B \times F \times G = D \times F \times J = E \times G \times J = C \times K$$

$$A \times J = C \times D \times E = C \times F \times G = D \times F \times H = E \times G \times H = B \times K$$

$$A \times K = E \times F = D \times G = C \times H = B \times J$$

$$B \times C = A \times D \times F = A \times E \times G = H \times J = D \times E \times K = F \times G \times K$$

$$B \times D = A \times C \times F = A \times E \times H = G \times J = C \times E \times K = F \times H \times K$$
$$B \times E = A \times C \times G = A \times D \times H = F \times J = C \times D \times K = G \times H \times K$$
$$B \times F = A \times C \times D = A \times G \times H = E \times J = C \times G \times K = D \times H \times K$$
$$B \times G = A \times C \times E = A \times F \times H = D \times J = C \times F \times K = E \times H \times K$$
$$B \times H = A \times D \times E = A \times F \times G = C \times J = D \times F \times K = E \times G \times K$$
$$C \times D = A \times B \times F = G \times H = A \times E \times J = B \times E \times K = F \times J \times K$$
$$C \times E = A \times B \times G = F \times H = A \times D \times J = B \times D \times K = G \times J \times K$$
$$C \times F = A \times B \times D = E \times H = A \times G \times J = B \times G \times K = D \times J \times K$$
$$C \times G = A \times B \times E = D \times H = A \times F \times J = B \times F \times K = E \times J \times K$$
$$D \times E = F \times G = A \times B \times H = A \times C \times J = B \times C \times K = H \times J \times K$$
$$D \times F = A \times B \times C = E \times G = A \times H \times J = B \times H \times K = C \times J \times K$$

For 11 factors in 32 trials main factors are clear of 2-factor interactions, but 2-factor interactions are confounded with one another. The aliases are

A = B × C × F = D × F × G = C × D × J = B × G × J = E × H × J = D × E × K = C × H × K = F × H × L = B × K × L

B = A × C × F = C × D × G = E × G × H = D × F × J = A × G × J = F × H × K = D × E × L = C × H × L = A × K × L

C = A × B × F = B × D × G = D × E × H = A × D × J = F × G × J = A × H × K = E × J × K = E × G × L = B × H × L = F × K × L

D = B × C × G = A × F × G = C × E × H = A × C × J = B × F × J = A × E × K =H × J × K = B × E × L = G × H × L

E = C × D × H = B × G × H − A × H × J = A × D × K = F × G × K = C × J × K = B × D × L = C × G × L = F × J × L

F = A × B × C = A × D × G = B × D × J = C × G × J = E × G × K = B × H × K = A × H × L = E × J × L = C × K × L

G = B × C × D = A × D × F = B × E × H = A × B × J = C × F × J = E × F × K = C × E × L = D × H × L = J × K × L

H = C × D × E = B × E × G = A × E × J = A × C × K = B × F × K = D × J × K = B × C × L = A × F × L = D × G × L

J = A × C × D = B × D × F = A × B × G = C × F × G = A × E × H = C × E × K = D × H × K = E × F × L = G × K × L

K = A × D × E = E × F × G = A × C × H = B × F × H = C × E × J = D × H × J = A × B × L = C × F × L = G × J × L

L = B × D × E = C × E × G = B × C × H = A × F × H = D × G × H = E × F × J = A × B × K = C × F × K = G × J × K

A × B = C × F = G × J = K × L   A × C = B × F = D × J = H × K

A × D = F × G = C × J = E × K   A × E = H × J = D × K

A × F = B × C = D × G = H × L   A × G = D × F = B × J

A × H = E × J = C × K = F × L   A × J = C × D = B × G = E × H

A × K = D × E = C × H = B × L   A × L = F × H = B × K

B × D = C × G = F × J = E × L   B × E = G × H = D × L

B × H = E × G = C × L = F × K   C × E = D × H = J × K = G × L

E × F = G × K = J × L

For 12 factors in 32 trials main factors are clear of 2-factor interactions, but 2-factor interactions are confounded with one another. The 2-factor aliases are

A × B = C × F = D × G = H × J = E × K   A × C = B × F = D × H = G × J = E × L

A × D = B × G = C × H = F × J = E × M   A × E = B × K = C × L = D × M

A × F = B × C = G × H = D × J = K × L   A × G = B × D = F × H = C × J = K × M

A × H = F × G = C × D = B × J = L × M   A × J = C × G = B × H = D × F

A × K = B × E = F × L = G × M   A × L = F × K = C × E = H × M

A × M = G × K = H × L = D × E   B × L = C × K = E × F = J × M

B × M = D × K = J × L = E × G   C × M = J × K = D × L = E × H

E × J = H × K = G × L = F × M   B × L = C × K = E × F = J × M

For 13 factors in 32 trials main factors are clear of 2-factor interactions, but 2-factor interactions are confounded with one another. The 2-factor aliases are

A × B = C × F = D × G = H × J = E × K = L × N   A × C = B × F = D × H = G × J = E × L = K × N

A × D = B × G = C × H = F × J = E × M   A × E = B × K = C × L = D × M = F × N

A × F = B × C = G × H = D × J = K × L = E × N   A × G = B × D = F × H = C × J = K × M

A × H = F × G = C × D = B × J = L × M   A × J = C × G = B × H = D × F = M × N

A × K = B × E = F × L = G × M = C × N   A × L = F × K = C × E = H × M = B × N

A × M = G × K = D × E = J × N   A × N = C × K = B × L = J × M = E × F

B × M = D × K = J × L = E × G = H × N   C × M = J × K = D × L = E × H = G × N

D × N = H × K = G × L = F × M = E × F

For 14 factors in 32 trials main factors are clear of 2-factor interactions, but 2-factor interactions are confounded with one another. The 2-factor aliases are

A × B = C × F = D × G = H × J = E × K = L × N = M × O

A × C = B × F = D × H = G × J = E × L = K × N   A × D = B × G = C × H = F × J = E × M = K × O

A × E = B × K = C × L = D × M = F × N = G × O   A × F = B × C = G × H = D × J = K × L = E × N

A × G = B × D = F × H = C × J = K × M = E × O   A × H = F × G = C × D = B × J = L × M = N × O

A × J = C × G = B × H = D × F = M × N = L × O   A × K = B × E = F × L = G × M = C × N = D × O

A × L = F × K = C × E = H × M = B × N = J × O   A × M = G × K = H × L = D × E = J × N = B × O

A × N = C × K = B × L = J × M = E × F = H × O   A × O = D × K = J × L = B × M = H × N = E × G

C × M = J × K = D × L = E × H = G × N = F × O   C × O = H × K = G × L = F × M = D × N = E × J

For 15 factors in 32 trials main factors are clear of 2-factor interactions, but 2-factor interactions are confounded with one another. The 2-factor aliases are

A × B = C × F = D × G = H × J = E × K = L × N = M × O

A × C = B × F = D × H = G × J = E × L = K × N = M × P   A × D = B × G = C × H = F × J = E × M = K × O = L × P

A × E = B × K = C × L = D × M = F × N = G × O = H × P   A × F = B × C = G × H = D × J = K × L = E × N = O × P

A × G = B × D = F × H = C × J = K × M = E × O = N × P   A × H = F × G = C × D = B × J = L × M = N × O = E × P

A × J = C × G = B × H = D × F = M × N = L × O = K × P   A × K = B × E = F × L = G × M = C × N = D × O = J × P

A × L = F × K = C × E = H × M = B × N = J × O = D × P   A × M = G × K = H × L = D × E = J × N = B × O = C × P

A × N = C × K = B × L = J × M = E × F = H × O = G × P   A × O = D × K = J × L = B = H × N = E × G = F × P

A × P = J × K = D × L = C × M = G × N = F × O = E × H   B × P = H × K = G × L = F × M = D × N = C × O = E × J

APPENDIX
D

Tables[1]

Table 1. *F* Distribution

					df_{x1}					
df_{x2}	1	2	3	4	5	6	7	8	9	10
0.10		39.9	49.5	53.6	55.8	57.2	58.2	58.9	59.4	59.9	60.2
0.05	1	161	200	216	225	230	234	237	239	241	242
0.01		4050	5000	5400	5620	5760	5860	5930	5980	6020	6060
0.10		8.53	9.00	9.16	9.24	9.29	9.33	9.35	9.37	9.38	9.39
0.05	2	18.5	19.0	19.2	19.2	19.3	19.3	19.4	19.4	19.4	19.4
0.01		98.5	99.0	99.2	99.2	99.3	99.3	99.4	99.4	99.4	99.4
0.10		5.54	5.46	5.39	5.34	5.31	5.28	5.27	5.25	5.24	5.23
0.05	3	10.1	9.55	9.28	9.12	9.01	8.94	8.89	8.85	8.81	8.79
0.01		34.1	30.8	29.5	28.7	28.2	27.9	27.7	27.5	27.3	27.2

(Continued)

[1] Reprinted with permission from the notes of "Product Quality and Process Efficiency," present by Dr. George Pouskouieli, Carlton Professional Development Centre.

Table 1. *F* Distribution (*Continued*)

0.10		4.54	4.32	4.19	4.11	4.05	4.01	3.98	3.95	3.93	3.92
0.05	4	7.71	6.94	6.59	6.39	6.26	6.16	6.09	6.04	6.00	5.96
0.01		21.2	18.0	16.7	16.0	15.5	15.2	15.0	14.8	14.7	14.5
0.10		4.06	3.78	3.62	3.52	3.45	3.40	3.37	3.34	3.32	3.30
0.05	5	6.61	5.79	5.41	5.19	5.05	4.95	4.88	4.82	4.77	4.74
0.01		16.3	13.3	12.1	11.4	11.0	10.7	10.5	10.3	10.2	10.1
0.10		3.78	3.46	3.29	3.18	3.11	3.05	3.01	2.98	2.96	2.94
0.05	6	5.99	5.14	4.76	4.53	4.39	4.28	4.21	4.15	4.10	4.06
0.01		13.7	10.9	9.78	9.15	8.75	8.47	8.26	8.10	7.98	7.87
0.10		3.59	3.26	3.07	2.96	2.88	2.83	2.78	2.75	2.72	2.70
0.05	7	5.59	4.74	4.35	4.12	3.97	3.87	3.79	3.73	3.68	3.64
0.01		12.2	9.55	8.45	7.85	7.46	7.19	6.99	6.84	6.72	6.62
0.10		3.46	3.11	2.92	2.81	2.73	2.67	2.62	2.59	2.56	2.54
0.05	8	5.32	4.46	4.07	3.84	3.69	3.58	3.50	3.44	3.39	3.35
0.01		11.3	8.65	7.59	7.01	6.63	6.37	6.18	6.03	5.91	5.81
0.10		3.36	3.01	2.81	2.69	2.61	2.55	2.51	2.47	2.44	2.42
0.05	9	5.12	4.26	3.86	3.63	3.48	3.37	3.29	3.23	(3.18)	3.14
0.01		10.6	8.02	6.99	6.42	6.06	5.8	5.61	5.47	5.35	5.26
0.10		3.29	2.92	2.73	2.61	2.52	2.46	2.41	2.38	2.35	2.32
0.05	10	4.96	4.10	3.71	3.48	3.33	3.22	3.14	3.07	3.02	(2.98)
0.01		10.0	7.56	6.55	5.99	5.64	5.39	5.2	5.06	4.94	4.85
0.10		3.18	2.81	2.61	2.48	2.39	2.33	2.28	2.24	2.21	2.19
0.05	12	4.75	3.89	3.49	3.26	3.11	3.00	2.91	2.85	2.80	2.75
0.01		9.33	6.93	5.95	5.41	5.06	4.82	4.64	4.50	4.39	4.30
0.10		3.07	2.70	2.49	2.36	2.27	2.21	2.16	2.12	2.09	2.06
0.05	15	4.54	3.68	3.29	3.06	2.90	2.79	2.71	2.64	2.59	2.54
0.01		8.68	6.36	5.42	4.89	4.56	4.32	4.14	4.00	3.89	3.80
0.10		2.97	2.59	2.38	2.25	2.16	2.09	2.04	2.00	1.96	1.94
0.05	20	4.35	3.49	3.10	2.87	2.71	2.60	2.51	2.45	2.39	2.35
0.01		8.10	5.85	4.94	4.43	4.10	3.87	3.70	3.56	3.46	3.37
0.10		2.88	2.49	2.28	2.14	2.05	1.98	1.93	1.88	1.85	1.82
0.05	30	4.17	3.32	2.92	2.69	2.53	2.42	2.33	2.27	2.21	2.16
0.01		7.56	5.39	4.51	4.02	3.70	3.47	3.30	3.17	3.07	2.98
0.10		2.79	2.39	2.18	2.04	1.95	1.87	1.82	1.77	1.74	1.71
0.05	60	4.00	3.15	2.76	2.53	2.37	2.25	2.17	2.10	2.04	1.99
0.01		7.08	4.98	4.13	3.65	3.34	3.12	2.95	2.82	2.72	2.63
0.10		2.75	2.35	2.13	1.99	1.90	1.82	1.77	1.72	1.68	1.65
0.05	120	3.92	3.07	2.68	2.45	2.29	2.18	2.09	2.02	1.96	1.91
0.01		6.85	4.79	3.95	3.48	3.17	2.96	2.79	2.66	2.56	2.47
0.10		2.71	2.30	2.08	1.94	1.85	1.77	1.72	1.67	1.63	1.60
0.05	∞	3.84	3.00	2.60	2.37	2.21	2.10	2.01	1.94	1.88	1.83
0.01		6.63	4.61	3.78	3.32	3.02	2.80	2.64	2.51	2.41	2.32

Table 1. *F* Distribution (*Continued*)

| | df_{x2} | \multicolumn{7}{c}{df_{x1}} |
		12	15	20	30	60	120	∞
0.10		60.7	61.2	61.7	62.3	62.8	63.1	63.3
0.05	1	244	246	248	250	252	253	254
0.01		6110	6160	6210	6260	6310	6340	6370
0.10		9.41	9.42	9.44	9.46	9.47	9.48	9.49
0.05	2	19.4	19.4	19.5	19.5	19.5	19.5	19.5
0.01		99.4	99.4	99.4	99.5	99.5	99.5	99.5
0.10		5.22	5.20	5.18	5.17	5.15	5.14	5.13
0.05	3	8.74	8.70	8.66	8.62	8.57	8.55	8.53
0.01		27.4	26.9	26.7	26.5	26.3	26.2	26.1
0.10		3.90	3.87	3.84	3.82	3.79	3.78	3.76
0.05	4	5.91	5.86	5.80	5.75	5.69	5.66	5.63
0.01		14.4	14.2	14.0	13.8	13.7	13.6	13.5
0.10		3.27	3.24	3.21	3.17	3.14	3.12	3.11
0.05	5	4.68	4.62	4.56	4.50	4.43	4.40	4.37
0.01		9.89	9.72	9.55	9.38	9.20	9.11	9.02
0.10		2.90	2.87	2.84	2.80	2.76	2.74	2.72
0.05	6	4.00	3.94	3.87	3.81	3.74	3.70	3.67
0.01		7.72	7.56	7.40	7.23	7.06	6.97	6.83
0.10		2.67	2.63	2.59	2.56	2.51	2.49	2.47
0.05	7	3.57	3.51	3.44	3.38	3.30	3.27	3.23
0.01		6.47	6.31	6.16	5.99	5.82	5.74	5.65
0.10		2.50	2.46	2.42	2.38	2.34	2.31	2.29
0.05	8	3.28	3.22	3.15	3.08	3.01	2.97	2.93
0.01		5.67	5.52	5.36	5.20	5.03	4.95	4.86
0.10		2.38	2.34	2.30	2.25	2.21	2.18	2.16
0.05	9	3.07	3.01	2.94	2.86	2.79	2.75	2.71
0.01		5.11	4.96	4.81	4.65	4.48	4.40	4.31
0.10		2.28	2.24	2.20	2.15	2.11	2.08	2.06
0.05	10	2.91	2.84	2.77	2.70	2.62	2.58	2.54
0.01		4.71	4.56	4.41	4.25	4.08	4.00	3.91
0.10		2.15	2.10	2.06	2.01	1.96	1.93	1.90
0.05	12	2.69	2.62	2.54	2.47	2.38	2.34	2.30
0.01		4.16	4.01	3.86	3.70	3.54	3.45	3.36
0.10		2.02	1.97	1.92	1.87	1.82	1.79	1.76
0.05	15	2.48	2.40	2.33	2.25	2.16	2.11	2.07
0.01		3.67	3.52	3.37	3.21	3.05	2.96	2.87
0.10		1.89	1.84	1.79	1.74	1.68	1.64	1.61
0.05	20	2.28	2.20	2.12	2.04	1.95	1.90	1.84
0.01		3.23	3.09	2.94	2.78	2.61	2.52	2.42
0.10		1.77	1.72	1.67	1.61	1.54	1.50	1.46
0.05	30	2.09	2.01	1.93	1.84	1.74	1.68	1.62
0.01		2.84	2.70	2.55	2.39	2.21	2.11	2.01

Table 1.   *F* Distribution (*Continued*)

0.10		1.66	1.60	1.54	1.48	1.40	1.35	1.29
0.05	60	1.92	1.84	1.75	1.65	1.53	1.47	1.39
0.01		2.50	2.35	2.20	2.03	1.84	1.73	1.60
0.10		1.60	1.54	1.48	1.41	1.32	1.26	1.19
0.05	120	1.83	1.75	1.66	1.55	1.43	1.35	1.25
0.01		2.34	2.19	2.03	1.86	1.66	1.53	1.38
0.10		1.55	1.49	1.42	1.34	1.24	1.17	1.00
0.05	∞	1.75	1.67	1.57	1.46	1.32	1.22	1.00
0.01		2.18	2.04	1.88	1.70	1.47	1.32	1.00

Table 2.   *t*-Distribution (Double-Sided Test)

df			P		
	$t0.0025$	$t0.005$	$t0.025$	$t0.050$
1	127	63.7	12.7	6.31
2	14.1	9.92	4.30	2.92
3	7.45	5.84	3.18	2.35
4	5.60	4.60	2.78	2.13
5	4.77	4.03	2.57	2.01
6	4.32	3.71	2.45	1.94
7	4.03	3.50	2.36	1.89
8	3.83	3.36	2.31	1.86
9	3.69	3.25	2.26	1.83
10	3.58	3.17	2.23	1.81
11	3.50	3.11	2.20	1.80
12	3.43	3.05	2.18	1.78
13	3.37	3.01	2.16	1.77
14	3.33	2.98	2.14	1.76
15	3.29	2.95	2.13	1.75
16	3.25	2.92	2.12	1.75
17	3.22	2.90	2.11	1.74
18	3.20	2.88	2.10	1.73
19	3.17	2.86	2.09	1.73
20	3.15	2.85	2.09	1.72
21	3.14	2.83	2.08	1.72
22	3.12	2.82	2.07	1.72
23	3.10	2.81	2.07	1.71
24	3.09	2.80	2.06	1.71
25	3.08	2.79	2.06	1.71
26	3.07	2.78	2.06	1.71
27	3.06	2.77	2.05	1.70
28	3.05	2.76	2.05	1.70
28	3.04	2.76	2.05	1.70

Table 2. *t*-Distribution (Double-Sided Test) (*Continued*)

30	3.03	2.75	2.04	1.70
40	2.97	2.70	2.02	1.68
60	2.91	2.66	2.00	1.67
120	2.86	2.62	1.98	1.66
∞	2.81	2.58	1.96	1.64

Table 3. Random Numbers[a]

33	49	9	64	26	43	5	48	17
7	14	48	17	32	52	46	54	36
41	58	39	31	20	10	4	26	3
46	60	46	58	45	59	45	53	40
16	11	16	46	27	18	5	60	37
17	12	33	53	48	7	57	6	39
20	23	34	55	51	32	48	20	41
30	17	55	37	27	62	14	52	45
31	35	52	18	50	19	27	65	33
16	5	6	17	41	10	36	29	25
13	5	31	57	49	52	53	65	42
26	40	28	52	47	60	45	19	17
23	21	58	4	41	2	19	62	49
39	6	62	19	14	19	2	6	25
17	58	8	61	61	53	23	56	6
2	14	1	24	8	54	42	55	29
3	9	10	40	64	38	49	7	11
46	19	57	54	44	10	34	19	40
27	23	13	38	40	36	58	56	12
15	38	64	33	62	33	5	51	45
63	34	20	21	30	41	49	44	11
55	7	17	30	48	35	15	8	48
17	4	8	39	13	51	16	19	51
13	18	54	11	48	62	7	37	44
8	25	49	47	48	12	23	47	16
59	34	19	13	33	45	27	15	4
58	64	6	52	38	63	59	36	8
16	49	47	40	44	13	38	2	22
54	6	59	16	44	46	38	8	34
58	22	15	47	49	34	6	39	20
15	10	31	24	32	37	15	61	27
8	6	59	5	12	62	33	50	61
31	59	31	2	34	11	41	40	25
22	11	4	50	55	33	63	41	10
30	35	27	51	32	22	53	54	57
10	64	42	25	34	54	21	32	8
23	29	64	37	48	21	18	29	34

[a]Random numbers generated using Borland's Quattro Pro® V4.0.

Table 4. Values for $t_{\alpha/2}$ and t_β at Selected Degrees of Freedom

df	$t_{0.100}$	$t_{0.050}$	$t_{0.025}$	$t_{0.010}$	$t_{0.005}$
1	3.078	6.314	12.706	31.821	63.657
5	1.476	2.015	2.571	3.365	4.032
10	1.372	1.812	2.228	2.764	3.169
15	1.341	1.753	2.131	2.602	2.947
20	1.325	1.725	2.086	2.528	2.845
25	1.316	1.708	2.060	2.485	2.787
∞	1.282	1.645	1.960	2.326	2.576

$$\alpha = 0.05$$

$$\frac{\alpha}{2} = 0.025$$

APPENDIX
E

References

American Society for Quality Control. Statistics Division, "Glossary and Tables for Statistical Quality Control." ASQC Quality Press, Milwaukee, WI, 1983.

Bandurek, G.R., Hughes, H.L., Crouch, D. The use of Taguchi methods in performance demonstrations. *Quality and Reliability Engineering,* Volume 6, pg. 121–131 1990.

Barker, A.M. Quality optimization by design: Global synergy. *ASQC Quality Congress Transactions,* 1990, pg. 1047–1050.

Barker, T.B. "Quality by Experimental Design." Marcel Dekker, New York, 1985.

Belaire, P.M., Deacon, R.J. "A Strategic Approach to Quality Improvement Using Design of Experiments Concepts and Methods." ASME Production Engineering Division, 1987.

Bendell, A., Disney, J., Baker, A.G. Taguchi methods—Comments on UK company experiences. *Proceedings of RELIABILITY '89,* June 14–16 1989. pg. 4C 1611–1615.

Bendell, A., Disney, J., Pridmore, W.A. "Taguchi Methods: Applications in World Industry." IFS Publications, London, 1989.

Box, G.E.P., Hunter, W.G., Hunter, J.S. "Statistics for Experimenters: An Introduction to Design, Data Analysis and Model Building." John Wiley, New York, 1978.

Bryce, G.R., Collette, D.R. Process characterization and optimization using statistical design techniques. *Proceedings of the Kodak Microelectronics Seminar INTERFACE '83,* November 1983, pg. 134–142.

Burgam, P.M. Design of experiments—The Taguchi way. *Manufacturing Engineering,* Vol. 94, No 5 May 1985, pg. 44–47.

Burr, J.T. The tools of quality, Part 1: Going with the flow (chart). *Quality Progress,* June 1990, pg. 64–67.

Byrne, D.M., Taguchi, S. The Taguchi approach to parameter design. *Quality Progress,* December 1987, pg. 19–26.

Davies, O.L. (editor). "The Design and Analysis of Industrial Experiments." Hafner, New York, 1954.

Deming, W.E. "Out of the Crisis." Massachusetts Institute of Technology Centre for Advanced Engineering Study, Cambridge, MA, 1986.

Diamond, W.A. "Practical Experiment Designs for Engineers and Scientists." Van Nostrand Reinhold, New York, 1981.

Fisher, R.A. "The Design of Experiments." Oliver and Boyd, London, 1942.

Freund, J.E. "Modern Elementary Statistics," 5th ed. Prentice-Hall, Eaglewood Cliffs, NJ, 1979.

Gunter, B. Statistically designed experiments, Part 2: The universal structure underlying experimentation. *Quality Progress,* February 1990, pg. 87–89.

Hahn, G.J. Some things engineers know about experimental design. *Journal of Quality Technology,* Vol. 9, No 1, January, 1977, pg. 13–20.

Hahn, G.J. Experimental design in the complex world. *Technometrics,* February 1984. pg. 19–31.

Hendrix, C.D. What every technologist should know about experimental design. *Chemtech,* March 1979. pg. 40–59.

Hogg, R.V., Boardman, T.J., Bryce, G.R. Statistics for engineers: A problem solving approach. *ASEE Annual Conference Proceedings,* 1987, pg. 1459–1470.

Ishikawa, K. "Guide to Quality Control." Asian Productivity Organization, Unipub, New York, 1982.

Jamieson, A. "Introduction to Quality Control." Reston Publishing Company, Reston, VA, 1982.

Joglekar, A.M., May, A.T. Product excellence through design of experiments. *Cereal Foods World,* Vol. 32, No 12, 1987. pg. 857–868.

Kackar, R.N. Off-line quality control parameter design, and the Taguchi method. *Journal of Quality Technology,* Vol. 17, No 4 October 1985, pg. 176–188.

Moroney, M.J. "Facts from Figures." Penguin, Baltimore, MD, 1951.

Nolan, T.W., Provost, L.P. Understanding variation. *Quality Progress,* May 1990, pg. 70–78.

Pfeifer, C.G. Planning efficient and effective experiments. *Manufacturing Engineering,* May 1988, pg. 35–39.

Phadke, M.S. "Quality Engineering Using Robust Design." Prentice-Hall, Englewood Cliffs, NJ, 1989.

Ross, P.J. "Taguchi Techniques for Quality Engineering." McGraw-Hill, New York, 1988.

→ Schleckser, J. "Use of Screening Experiments to Optimize Product Performance and Quality." Promotional Literature, Rogers Corporation, Manchester, CT, 1986.

→ Schmidt, S.R., Launsby, R.G. "Understanding Industrial Designed Experiments." Air Academy Press, Colorado Springs, CO, 1988.

Schultz, L.E., Schroeder, D.R. "Pathway to Continuous Process Improvement." Promotional Literature, Process Management Institute, Minneapolis, MN. 1986.

Shewhart, W.A. "Economic Control of Quality of Manufactured Product." Van Norstrand, New York, 1931.

— Snee, R.D., Hare, L.B., Trout, J.R. "Experiments in Industry: Design, Analysis and Interpretation of Results." ASQC Quality Press, Milwaukee, WI, 1985.

→ Taguchi, G. "System of Experimental Design" (Volumes 1 and 2). UNIPUB/Kraus International Publications, White Plains, NY and American Supplier Institute, Dearborn, MI, 1987.

→ Tsui, Kwok-Leung. Strategies for planning experiments using orthogonal arrays and confounding tables. *Quality and Reliability Engineering, Volume 4, 1988, pg. 113–122.*

— Wadsworth, H.M. "Handbook of Statistical Methods for Engineers and Scientists." McGraw-Hill, New York, 1990.

Wadsworth, H.M., Stephens, K.S., Godfrey, A.B. "Modern Methods for Quality Control and Improvement." John Wiley, New York, 1986.

→ Wheeler, D.J. "Understanding Industrial Experimentation." Statistical Process Controls, Inc., Knoxville, TN, 1988.

Wheeler, D.J., Chambers, D. A. "Understanding Statistical Process Control." Statistical Process Controls, Inc., Knoxville, TN, 1986.

— Youden, W.J. "Statistical Methods for Chemists." John Wiley, New York, 1951.

Young, J.C. Experimental design: A strategy for implementation. *ASQC Quality Congress Transactions,* 1989, pg. 34–40.

Index

Action step, 12
Alpha (α)-risk, 57
Alias structure (*see* Confounding)
Alternate hypothesis, 56
  also see Hypothesis
Analysis of variance
  definition of, 61
  table,
    blank, 66
    baseball experiment, 99
    fuel efficiency example 1, 68
    fuel efficiency example 2, 75
ANOVA, 61 (*see also* Analysis of variance)
Average
  definition of, 45
  formula, 45
  weakness, 46

Best run conditions
  estimation of,
    baseball experiment, 100
    formula for, 100
Brainstorming, 6
  guidelines, 39

Controllable factor (*see* Factor)
Case study
  baseball experiment, 89
Cause and effect, 6
  alternate construction method, 41
  benefits, 33
  case enumeration, 37
  construction of, 37
  definition of, 33
  dispersion analysis, 35
  identifying subcauses, 79
  key factors, 37
  number of categories, 39
  other uses, 35
  prioritizing causes, 40
  process classification, 35
  required number of causes, 40
  selecting causes, 40
  short lists, 40
  types of diagrams, 35
Change-one-factor-at-a-time, 3

Check sheet
  categorizing data, 22
  example of,
    fuel efficiency, 63
  types of, 15
  construction of check sheets, 16
    collecting data, 18
    collection points, 18
    objectives, 18
    review, 19
    time limits, 18
  uses of, 16
Confidence interval
  baseball experiment, 101
  definition of, 101
  error estimate,
    formula for, 101
Confidence level, 56
Confirmation experiment, 102
Confounding
  alias structure, 79
  definition of, 78
  development of, 78
  for common arrays, 81
Correction factor
  definition of, 64
  baseball experiment, 95
  fuel efficiency, example 1, 67
  fuel efficiency, example 2, 73
Cross product, 81
Customer, 5

Decision step, 12
Degrees of freedom
  definition of, 64
  calculation of
    baseball experiment, 97
    fuel efficiency, example 1, 68
    fuel efficiency, example 2, 74
Denominator, 64
Designed experimentation
  design of experiments, 2
  DoE, 2
  statistically designed experiments, 2
Design selection
  criteria, 82

resolution, 82

80/20 rule, 25
Error
    Type I, 56
    Type II, 56
Expected sum of squares
    definition of, 98
    formula for, 98

Factor
    definition of, 33
    ranges
        selection of, 85
Flow diagrams
    consistency, 14
    construction of, 10
    guidelines, 10
    secondary loop, 13
    list of
        ANOVA calculation, 91
        change-one-factor-at-a-time, 3
        cause and effect diagrams, 38
        check sheet, 19
        Pareto charts, 27
        pretrial planning, 7
        stages of experimentation, 4
        symbols, 12
        words, 10
F statistic
    calculation of
        baseball experiment, 97
        fuel efficiency, example 1, 69
        fuel efficiency, example 2, 75
    determination of, 57
    formula, 57
    experimental, 57
    theoretical, 57
Full factorial experiment
    definition of, 82

Group 1 factors, 84
Group 2 factors, 84

Hypothesis
    accepting or rejecting, 57
    alternate, 56
    example of,
        fuel efficiency, example 1, 62

fuel efficiency, example 2, 71
    formulating, 54
    null, 54

Ina Seito
    INAX, 3
Interactions (see also Confounding)
    complexity, 80
    definition, 79
    location, 80

Level sums, 91
    calculation check, 93
    table of, 93

Mean
    definition of, 45
    formula, 46
Median
    definition of, 46
    formula, 47
Mean sum of squares, 68

Normal distribution, 53
Null hypothesis (also see Hypothesis)
Numerator, 64

One-way analysis of variance
    blank table of, 66
    definition of, 61
    squares calculation
        fuel efficiency, example 1, 66
        fuel efficiency, example 2, 71
    methodology, 61
Orthogonal arrays
    balanced designs, 78
    column assignment, 84
    common designs, 72
    definition of, 73
    L_4, 78
    L_8, 79

Pareto charts
    blank template, 30
    construction of, 27
        categorizing data, 28
        plotting data, 30
        relative frequency, 29
        selecting a format, 28

cumulative frequency, 29
data order, 26
definition, 25
strengths, 26
Percent contribution
calculation of baseball experiment, 98
definition of, 98
formula for, 98
Plackett–Burman designs, 77
Pooled error
definition of, 96
degrees of freedom, 97
Population
definition of, 44
standard deviation, 49

Quality characteristic, 6, 83
definition of, 33

Randomization, 84
Resolution
definition of, 82
types of, 82
Response variable
definition, 83

Sample
definition of, 44
sizes
determination of, 86
example, 87
formula used, 87
improvements to precision, 86
number of samples required, 86
standard deviation
alternate formula, 52
definition of, 49
formula, 50
Sample variance
definition of, 49
formula, 50
Saturated designs, 82
Screening designs
types of, 77
fractional factorial designs, 77
orthogonal arrays, 77
Significance plots, 96
Simple tools
check sheets, 15
Pareto charts, 25

cause and effect, 33
flow diagrams, 9
Stages of experimentation
Stage 1
design team, 4
opportunity identification, 6
Stage 2
preparation for experiment, 6
pretrial preparation, 7
table of reasonableness, 7
Stage 3
conduct trials, 7
Stage 4
data analysis, 8
Stage 5
publicize results, 9
Standard deviation
alternate formula, 52
definition of, 49
population
formula, 50
sample
formula, 50
Statistics
definition of, 43
descriptive, 44
Sums of squares, 64
calculation of
baseball experiment, 94
fuel efficiency, example 1, 67
fuel efficiency, example 2, 73
definition of, 64

Table of Reasonableness
definition, 7
example of
baseball experiment, 93
epoxy experiment, 86
Treatment, 65
Two-way analysis of variance
definition of, 70

Unsaturated designs
definition of, 82

Variance, 49, 64, 97
Variation
between, 63
total, 63
within, 63